BIODIVERSITY MONITORING IN AUSTRALIA

Editors:
David Lindenmayer and Philip Gibbons
The Australian National University

CSIRO

PUBLISHING

National Library of Australia Cataloguing-in-Publication entry

Biodiversity monitoring in Australia/edited by David Lindenmayer and Philip Gibbons.

9780643103573 (pbk.)
9780643103580 (epdf)
9780643103597 (epub)

Includes bibliographical references and index.

Biodiversity – Monitoring – Australia.

Lindenmayer, David.
Gibbons, Philip.

333.95110994

Published by
CSIRO PUBLISHING
36 Gardiner Road, Clayton VIC 3168
Private Bag 10, Clayton South VIC 3169
Australia

Telephone: [+613] 9545 8555
Local call: 1300 788 000 (Australia only)
Fax: +61 3 9662 7555
Email: csiropublishing@csiro.au
Web site: www.publishing.csiro.au

Front cover image by iStockphoto

Set in 10/12 Adobe Minion Pro and ITC Stone Sans
Edited by Anne Findlay, Editing Works Pty Ltd
Cover and text design by James Kelly
Typeset by Desktop Concepts Pty Ltd, Melbourne
Index by Bruce Gillespie
Printed by Ingram Lightning Source

CSIRO PUBLISHING publishes and distributes scientific, technical and health science books and journals from Australia to a worldwide audience and conducts these activities autonomously from the research activities of the Commonwealth Scientific and Industrial Research Organisation (CSIRO). The views expressed in this publication are those of the author(s) and do not necessarily represent those of, and should not be attributed to, the publisher or CSIRO.

Contents

Acknowledgements   *vii*
List of contributors   *ix*

**Introduction: making monitoring happen – and then delivering on
Australia's Biodiversity Conservation Strategy**   1
David Lindenmayer and Philip Gibbons

National and international perspectives   **5**

1   Making monitoring up-front and centre in Australian biodiversity
    conservation   7
    David Lindenmayer

2   Accountability: we're an indulgent and marginal profession if
    we can't measure the effectiveness of investment in environmental
    management   15
    John C.Z. Woinarski

3   Ecoinformatics solutions to support monitoring for improved
    biodiversity conservation   23
    Andre Zerger and Warwick McDonald

4   Monitoring Australian birds to meet international obligations   33
    Stephen T. Garnett

5   Cheerfulness and grumpiness in ecological monitoring in Australia 43
    Richard J. Hobbs

6   The conservation return on investment from ecological monitoring   49
    *Hugh P. Possingham, Brendan A. Wintle, Richard A. Fuller and
    Liana N. Joseph*

7   Big-picture assessment of biodiversity change: scaling up
    monitoring without selling out on scientific rigour   63
    Simon Ferrier

8   An endpoint hierachy and process control charts for ecological
    monitoring   71
    *Mark Burgman, Kim Lowell, Peter Woodgate, Simon Jones, Gary Richards
    and Prue Addison*

9   Lessons from environment accounting for improving biodiversity
     monitoring   79
     Michael Vardon

Government agency and NGO perspectives   89

10   Cows, cockies and atlases: use and abuse of biodiversity
     monitoring in environmental decision-making   91
     James Fitzsimons

11   A park manager's perspective on ecological monitoring   101
     Tony Varcoe

12   Monitoring for improved biodiversity outcomes in the private
     conservation estate: perspective from Bush Heritage Australia   111
     *Jim Radford, Murray Haseler, Sandy Gilmore, Angela Sanders, Adam
     Kerezsy, Max Tischler and Matthew Appleby*

13   Practical challenges in monitoring and adapting restoration
     strategies and actions   121
     David Freudenberger

14   Measuring and reporting on conservation management outcomes   127
     Sarah Legge and Atticus Fleming

15   Making monitoring work for conservation: lessons from
     The Nature Conservancy   133
     Jensen R. Montambault and Craig Groves

16   Biodiversity monitoring from a community organisation perspective 141
     Doug Robinson, Lisa Smallbone and James O'Connor

Programs and the lessons learned from them   149

17   Biodiversity monitoring in Canada's Yukon: The Community
     Ecological Monitoring Program   151
     Charles J. Krebs

18   Monitoring for improved biodiversity conservation in arid
     Australia   157
     Chris R. Dickman and Glenda M. Wardle

19   Exploiting the back-loop of the adaptive cycle: lessons from the
     Black Saturday Fires   165
     Philip Gibbons

20   Waterbird monitoring in Australia: value, challenges and
      lessons learnt after more than 25 years   171
      Richard T. Kingsford and John L. Porter

21   Biodiversity monitoring in the Australian rangelands   179
      I. Andrew White, Jeff N. Foulkes, Ben D. Sparrow and Andrew J. Lowe

Discussion   191

22   Can we make biodiversity monitoring happen in Australia? Moving
      beyond 'It's the thought that counts'   193
      David Lindenmayer and Philip Gibbons

      Appendix 1: the workshop design, data capture and recording process   *202*
      Index   *204*

Acknowledgements

The February 2011 meeting that gave rise to this book was the brainchild of Max Bourke. The meeting was kindly funded by The Thomas Foundation and we give sincere thanks to David Thomas for financial, logistical and intellectual support.

Steve Colman was responsible for the creation and application of the iMEET! technology that fostered the capture of many of the insights and ideas that arose from the broad-ranging discussions that characterised the two days of meetings about biodiversity conservation.

Claire Shepherd, Jane Carter and Michelle Yin assisted with many of the organisational aspects of the meeting that ensured that the participants were focused on tackling the plethora of issues associated with biodiversity monitoring in Australia.

We are most grateful to all of the chapter authors of this book, who wrote and re-wrote chapters about biodiversity monitoring, took time out of their busy schedules to attend the meeting and who remain passionate about arresting the decline of biodiversity in Australia.

Finally, we thank John Manger and the staff at CSIRO Publishing for their hard work and encouragement in turning ideas and chapter contributions into a book.

David Lindenmayer and Phil Gibbons

List of contributors

Prue Addison
Australian Centre of Excellence for Risk Analysis, School of Botany, University of Melbourne, Melbourne, Vic 3010, Australia. Email: p.addison@student.unimelb.edu.au

Matthew Appleby
Bush Heritage Australia, Level 5, 395 Collins St., Melbourne, Vic 3000, Australia. Email: mappleby@bushheritage.org.au

Mark Burgman
Australian Centre of Excellence for Risk Analysis, School of Botany, University of Melbourne, Melbourne, Vic 3010, Australia. Email: markab@unimelb.edu.au

Chris R. Dickman
Desert Ecology Research Group, School of Biological Sciences, University of Sydney, Sydney, NSW 2006, Australia. Email: chris.dickman@sydney.edu.au

Simon Ferrier
CSIRO Ecosystem Sciences, GPO Box 1700, ACT 2601, Australia. Email: simon.ferrier@csiro.au

James Fitzsimons
The Nature Conservancy, Suite 3-04, 60 Leicester St., Carlton, Vic 3053, Australia. Email: jfitzsimons@tnc.org; and School of Life and Environmental Sciences, Deakin University, 221 Burwood Highway, Burwood, Vic 3125, Australia. Email: james.fitzsimons@deakin.edu.au

Atticus Fleming
Australian Wildlife Conservancy, PO Box 8070, Subiaco East, WA 6008, Australia. Email: atticus.fleming@australianwildlife.org

Jeff Foulkes
Terrestrial Ecosystem Research Network (AusPlots Facility), Australian Centre for Evolutionary Biology and Biodiversity, University of Adelaide, North Terrace, Adelaide, SA 5005, Australia; and Science Resource Centre and State Herbarium of South Australia, Plant Biodiversity Centre, South Australian Department of Environment and Natural Resources, Adelaide, SA 5005, Australia. Email: jeff.foulkes@adelaide.edu.au

David Freudenberger
Greening Australia, PO Box 74, Yarralumla, ACT 2600, Australia. Email: dfreudenberger@greeningaustralia.org.au

Richard A. Fuller
School of Biological Sciences, The University of Queensland, St Lucia, Qld 4072, Australia; and CSIRO Climate Adaptation Flagship and CSIRO Ecosystem Sciences, St Lucia, Qld 4072, Australia. Email: r.a.fuller@dunelm.org.uk

Stephen T. Garnett
Research Institute for the Environment and Livelihoods, Charles Darwin University, Darwin, NT 0909, Australia. Email: stephen.garnett@cdu.edu.au

Philip Gibbons
Fenner School of Environment and Society, The Australian National University, Canberra, ACT 0200, Australia. Email: philip.gibbons@anu.edu.au

Sandy Gilmore
Bush Heritage Australia, Level 5, 395 Collins St., Melbourne, Vic 3000, Australia. Email: sgilmore@bushheritage.org.au

Craig Groves
The Nature Conservancy, 40 East Main St., Suite 200, Bozeman, Montana 59715, USA. Email: craig_groves@tnc.org

Murray Haseler
Bush Heritage Australia, Level 5, 395 Collins St., Melbourne, Vic 3000, Australia. Email: mhaseler@bushheritage.org.au

Richard J. Hobbs
School of Plant Biology, University of Western Australia, Crawley, WA 6009, Australia. Email: richard.hobbs@uwa.edu.au

Simon Jones
School of Mathematical and Geospatial Sciences, RMIT University, Melbourne, Vic 3000, Australia. Email: simon.jones@rmit.edu.au

Liana N. Joseph
Wildlife Conservation Society, 2300 Southern Boulevard, The Bronx, New York 10460, USA. Email: ljoseph@wcs.org

Adam Kerezsy
Bush Heritage Australia, Level 5, 395 Collins St., Melbourne, Vic 3000, Australia. Email: akerezsy@bushheritage.org.au

Richard T. Kingsford
Australian Wetlands and Rivers Centre, School of Biological, Earth and Environmental Sciences, University of New South Wales, Sydney, NSW 2052, Australia. Email: richard.kingsford@unsw.edu.au

Charles J. Krebs

Department of Zoology, University of British Columbia, Vancouver, BC V6T 1Z4, Canada; and Institute for Applied Ecology, University of Canberra, Canberra, ACT 2601, Australia. Email: krebs@zoology.ubc.ca

Sarah Legge

Australian Wildlife Conservancy, Mornington Wildlife Sanctuary, PMB 925, Derby, WA 6728, Australia. Email: sarah@australianwildlife.org

David Lindenmayer

Fenner School of Environment and Society, and ARC Centre of Excellence for Environmental Decisions, The Australian National University, Canberra, ACT 0200, Australia. Email: david.lindenmayer@anu.edu.au

Andrew Lowe

Terrestrial Ecosystem Research Network (AusPlots Facility), Australian Centre for Evolutionary Biology and Biodiversity, University of Adelaide, North Terrace, Adelaide, SA 5005, Australia; and Science Resource Centre and State Herbarium of South Australia, Plant Biodiversity Centre, South Australian Department of Environment and Natural Resources, Adelaide, SA 5005, Australia. Email: andrew.lowe@adelaide.edu.au

Kim Lowell

Cooperative Research Centre for Spatial Information, 723 Swanston St., Ground floor, Carlton, Vic 3052, Australia. Email: klowell@unimelb.edu.au

Warwick McDonald

Environmental Information Services, Bureau of Meteorology, GPO Box 2334, Canberra, ACT 2600, Australia. Email: w.mcdonald@bom.gov.au

Jensen R. Montambault

The Nature Conservancy, 490 Westfield Road, Charlottesville, Virginia 22901, USA. Email: jmontambault@tnc.org

James O'Connor

Birds Australia, Suite 2-05, 60 Leicester St., Carlton, Vic 3053, Australia. Email: j.oconnor@birdsaustralia.com.au

John L. Porter

Australian Wetlands and Rivers Centre, School of Biological, Earth and Environmental Sciences, University of New South Wales, Sydney, NSW 2052, Australia. Email: john.porter@unsw.edu.au

Hugh P. Possingham

The University of Queensland, The Ecology Centre, School of Biological Sciences, and ARC Centre of Excellence for Environmental Decisions, St Lucia, Qld 4072, Australia. Email: h.possingham@uq.edu.au

Jim Radford
Bush Heritage Australia, Level 5, 395 Collins St., Melbourne, Vic 3000, Australia. Email: jradford@bushheritage.org.au

Gary Richards
Fenner School of Environment and Society, The Australian National University, Canberra, ACT 0200, Australia. Email: gary.richards@climatechange.gov.au

Doug Robinson
Trust for Nature, Level 5/379 Collins St., Melbourne, Vic 3000, Australia. Email: dougr@tfn.org.au

Angela Sanders
Bush Heritage Australia, Level 5, 395 Collins St., Melbourne, Vic 3000, Australia. Email: asanders@bushheritage.org.au

Lisa Smallbone
Trust for Nature, Level 5/379 Collins St., Melbourne, Vic 3000; and School of Environmental Sciences, Charles Sturt University, PO Box 789, Albury, NSW 2640, Australia. Email: lsmallbone@csu.edu.au

Ben Sparrow
Terrestrial Ecosystem Research Network (AusPlots Facility), Australian Centre for Evolutionary Biology and Biodiversity, University of Adelaide, North Terrace, Adelaide, SA 5005, Australia. Email: ben.sparrow@adelaide.edu.au

Max Tischler
Bush Heritage Australia, Level 5, 395 Collins St., Melbourne, Vic 3000, Australia. Email: mtischler@bushheritage.org.au

Tony Varcoe
Parks Victoria, Level 10, 535 Bourke St., Melbourne, Vic 3000, Australia. Email: tony.varcoe@parks.vic.gov.au

Michael Vardon
Centre of Environment and Energy Statistics, Australian Bureau of Statistics, 45 Benjamin Way, Belconnen, ACT 2616, Australia. Email: michael.vardon@abs.gov.au

Glenda M. Wardle
Desert Ecology Research Group, School of Biological Sciences, University of Sydney, Sydney, NSW 2006, Australia. Email: glenda.wardle@sydney.edu.au

I. Andrew White
Terrestrial Ecosystem Research Network (AusPlots Facility), Australian Centre for Evolutionary Biology and Biodiversity, University of Adelaide, North Terrace, Adelaide, SA 5005, Australia. Email: andrew.white@adelaide.edu.au

Brendan A. Wintle
School of Botany, The University of Melbourne, Parkville, Melbourne, Vic 3010, Australia; and ARC Centre of Excellence for Environmental Decisions, The University of Melbourne, Parkville, Melbourne, Vic 3010, Australia. Email: brendanw@unimelb.edu.au

John C.Z. Woinarski
Research Institute for the Environment and Livelihoods, Charles Darwin University, Darwin, NT 0909, Australia. Email: john.woinarski@cdu.edu.au

Peter Woodgate
Cooperative Research Centre for Spatial Information, 723 Swanston St., Ground floor, Carlton, Vic 3052, Australia. Email: pwoodgate@crcsi.com.au

Andre Zerger
Environmental Information Services, Bureau of Meteorology, GPO Box 2334, Canberra, ACT 2600, Australia. Email: a.zerger@bom.gov.au

INTRODUCTION: MAKING MONITORING HAPPEN – AND THEN DELIVERING ON AUSTRALIA'S BIODIVERSITY CONSERVATION STRATEGY

David Lindenmayer and Philip Gibbons

Introduction

Australia is one of the world's few mega-diverse nations. It also has an appalling record of loss and decline in that mega-diversity. This is despite the billions of dollars that are invested in biodiversity conservation annually. As of 2003/04, the total annual governmental expenditure on environmental management exceeded $12 billion per year (Beeton *et al.* 2006). It is likely to be far higher now. Moreover, repeated reports by the Australian National Audit Office (ANAO 1997; 2008) have indicated that it is not possible to assess the effectiveness of some of the largest conservation programs in this nation (e.g. the Natural Heritage Trust (Hajkowicz 2009)). Similarly, it is not possible to report back to the Australian taxpayer on the return for their investment in conservation programs.

A paucity of high-quality monitoring lies at the heart of the inability to assess effectiveness of conservation efforts, assess conservation return on investment, and ultimately to effectively stem (or reverse) the loss of biodiversity in Australia. This book tackles many aspects of the vexed problem of biodiversity monitoring. It arose from a major workshop held at The Australian National University in February 2011. The meeting was attended by leaders in the science, policy-making and management arenas of biodiversity conservation, and who have a particular interest, experience and expertise in biodiversity monitoring. The participants at the meeting were from all of Australia's states and territories and represented different federal, state and local government bodies as well as a number of national and international non-government organisations. There also were scientists from a number of Australian universities. Several international experts also participated. This diversity of participants was deliberate – successful biodiversity monitoring is dependent on partnerships among people with different kinds of expertise.

Chapter structure

Participants at the February 2011 workshop were asked to write their chapters prior to the meeting. This was done for three reasons. First, it ensured they had thought deeply about the issues associated with the poor record of biodiversity monitoring in Australia and how such problems might be resolved. Second, it increased the quality of the dialogue and the creativity of ideas at the meeting. Third, we decided at the early planning stages of the meeting that the ideas and insights about biodiversity monitoring should be captured in a readily accessible book. This was because many past discussions and reviews about biodiversity monitoring in Australia have either never been written up or were 'quasi-published' in documents that are difficult (if not impossible) to obtain. We argue that it is essential to record insights and recommendations on biodiversity monitoring so that others in the future can assess the progress that has (or has not) been made.

Each chapter author was provided with a template to guide the writing of their chapter. This has given rise to the broadly consistent chapter structure that is characteristic of all the contributions in this book. Specifically, we asked each chapter author to outline 2–3 successes in biodiversity monitoring, 2–3 problems and/or failures in biodiversity monitoring, and 3–4 solutions to current problems. The key themes associated with each success, failure and solution are summarised in the box at the beginning of each chapter. Some authors framed their writing around key statements in the recently released *Australia's Biodiversity Conservation Strategy 2010–2030* (Natural Resource Management Ministerial Council 2010) including:

> 'By 2015, establish a national long-term biodiversity monitoring and reporting system.
>
> Resources available for biodiversity conservation efforts—human and financial, government and non-government—are limited. It is therefore essential that we measure, evaluate and understand the effectiveness of our biodiversity conservation efforts. This knowledge will help to ensure that our efforts are correctly prioritised and targeted, so that we are investing in efficient actions that will produce the greatest long-term benefits for biodiversity.
>
> In order to get measurable results, we need to improve and share our knowledge of biodiversity. This involves improving the accessibility, communication and application of knowledge as well as ensuring our priorities are evidence-based. To achieve that, we need to implement robust national monitoring, reporting and evaluation measures, so that we can identify what is working—and not working—and why, and adjust our efforts accordingly.'

Given such statements, one of the subsidiary aims of this book was to propose approaches that will help deliver on the objectives set out above and others contained in *Australia's Biodiversity Conservation Strategy 2010–2030* (Natural Resource Management Ministerial Council 2010).

The structure of this book

An initial concern with commissioning chapters for this book was that there would be enormous overlap in the contributions and we would end up with 20 chapters with similar content and calling for the same set of recommendations. This did not eventuate. In fact, the content of the chapters is as diverse as it is insightful and thought provoking. Indeed, the diversity of the chapters made it difficult to determine an appropriate order of appearance for the book. After much thought, we elected to present the chapters as follows. The first set of chapters is broadly (but also loosely) grouped under the theme of national and international perspectives. The second grouping corresponds broadly to those chapters that have examined biodiversity monitoring from the perspective of a government agency or non-government organisation. The third set of chapters concerns individual monitoring programs or particular methods to improve monitoring and the lessons that have arisen from them.

The final part of this book is a general discussion. It is built around two components: (1) A general overview of the different perspectives that characterise the chapters in the preceding three parts of the book – a standard part of most concluding chapters in any edited book. (2) A synthesis of material derived from the extensive discussions at the meeting in February 2011.

Part of this synthesis was generated from a novel computer-based collation of problems thwarting the establishment of more and better biodiversity monitoring programs as well as solutions to these problems.

Software called iMEET! (http://imeet.com.au) was used to capture observations, problems and solutions throughout the meeting. We have briefly outlined this process in Appendix 1 at the end of this volume.

Some caveats

We are acutely aware that only a small subset of those people with expertise in biodiversity monitoring was invited to attend our February 2011 meeting and contribute a chapter to this book. However, the meeting was necessarily small to allow sensible and tractable discussions (presenters of talks at the meeting could not exceed 5 minutes). Given this, we fully acknowledge that there will be other perspectives on biodiversity monitoring that have not been represented, either in part or in full, in this volume. In many respects this is good because it means there is a lot more to be said and written about biodiversity monitoring for improved conservation outcomes. If this book can stimulate additional dialogue and, in turn, foster more support for more and better biodiversity monitoring, then we strongly believe that this exercise will have been a valuable one.

Overarching aims

This book has three aims. The first is to canvass opinion of what has led to successful monitoring, what the key problems with biodiversity monitoring are, and what the practical solutions to those problems should be. By capturing critical insights into successes, failures and solutions, we hope to provide high-level guidance for important initiatives like the Biodiversity Conservation Strategy, similar kinds of conservation initiatives in state government agencies, as well as non-government organisations that aim to improve conservation outcomes in Australia. We also hope to considerably improve the quality and effectiveness of biodiversity monitoring in Australia. Ultimately, our third aim, which follows from the first two aims, is to arrest the decline of biodiversity in Australia.

David Lindenmayer and Philip Gibbons
2011

References

ANAO (1997). 'Commonwealth natural resource management and environment programs'. Report No. 36, 1996–97. Performance Audit. The Australian National Audit Office, Canberra.

ANAO (2008). 'Regional delivery model for the Natural Heritage Trust and the National Action Plan for Salinity and Water Quality'. Report No. 21, 2007–08, Performance Audit. On-line: <http://www.anao.gov.au/download.cfm?item_id=CE21 C4471560A6E8AA2773F E12C20785&binary_id=F20BFC111560A6E8AA75D180CD56E65 2> [Accessed 10/02/2011] Australian National Audit Office, Canberra.

Beeton RJS, Buckley KI, Jones GJ, Morgan D, Reichelt RE and Trewin D (2006). 'Australia state of the environment'. Independent report to the Australian Government Minister for the Environment and Heritage. Department of the Environment and Heritage, Canberra.

<http://www.environment.gov.au/soe/2006/publications/report/pubs/soe-2006-report.pdf>

Hajkowicz S (2009). The evolution of Australia's natural resource management programs: Towards improved targeting and evaluation of investments. *Land Use Policy* **26**, 471–478.

Natural Resource Management Ministerial Council (2010). 'Australia's Biodiversity Conservation Strategy 2010–2030'. Australian Government Department of Sustainability, Environment, Water, Population and Communities, Canberra. <http://www.environment.gov.au/biodiversity/strategy/index.html>

NATIONAL AND INTERNATIONAL PERSPECTIVES

1 MAKING MONITORING UP-FRONT AND CENTRE IN AUSTRALIAN BIODIVERSITY CONSERVATION

David Lindenmayer

Lesson #1. Monitoring has highlighted some outstanding monitoring successes like the recovery of the Grey-crowned Babbler after management interventions in Victoria.

Lesson #2. It is now possible to identify some simple attributes characteristic of successful monitoring programs.

Lesson #3. There is an increased recognition among government and non-government agencies of the importance of monitoring.

Lesson #4. There are far too few well-designed and implemented long-term biodiversity monitoring studies in Australia.

Lesson #5. There has been a demonstrable failure to slow biodiversity decline in Australia.

Lesson #6. There has been a failure by a large proportion of the scientific fraternity to engage in monitoring.

Lesson #7. New forms of long-term funding arrangements must be developed to support high quality, biodiversity monitoring programs in Australia.

Lesson #8. New institutional arrangements are urgently required to underpin effective biodiversity monitoring programs in Australia.

Lesson #9. New approaches are needed to link question-driven with mandated biodiversity monitoring programs.

Introduction

Australia has an appalling record on biodiversity monitoring. While there have been some outstanding monitoring successes, it seems that these are more than countered by a long list of failed programs that remain uncompleted, or poor quality programs that are characterised by many problems. As an example, Allen (1993) and Norton (1996) described how nearly half of the more than 55 monitoring programs on tussock grasslands in New Zealand were unreported, indicating a failure rate that is extremely high. These problems also characterise many biodiversity monitoring programs in Australia. Yet, high quality monitoring is critical for many reasons including (among others):

1. Guiding effective management interventions.
2. Gauging progress in conservation programs and providing a basis for assessing return on conservation investment.
3. Highlighting examples of conservation success as an antidote to the otherwise overwhelming number of negative stories in biodiversity conservation (Garnett and Lindenmayer 2011) and, in turn, building ongoing (and hopefully increased) public and political support for conservation programs.

These (and many other) reasons mean that it is critical to significantly increase the number and quality of biodiversity monitoring programs in Australia. Based on my experience in long-term ecological work in south-eastern Australia over the past three decades, I outline my perspectives on monitoring successes and failures as well as what I strongly believe is needed to improve biodiversity monitoring in this country.

Successes

1. Monitoring has highlighted some outstanding monitoring successes like the recovery of the Grey-crowned Babbler after management interventions in Victoria

Conservation biology is widely regarded as a crisis discipline and, like many efforts in crisis management, failures are common. Much of conservation biology has therefore become a negative pursuit in which successes are rare. However, Garnett and Lindenmayer (2011) argue that more evidence of success is needed in conservation biology to provide motivation for redoubled efforts and tangible signals of positive results that arise from appropriate management interventions. Well-designed and implemented monitoring is fundamental to demonstrating conservation success and there are some good examples in an Australian biodiversity conservation context. For example, Robinson (2010) highlighted how management efforts on private agricultural land in Victoria designed to enhance the conservation of the threatened bird species, the Grey-crowned Babbler (*Pomatostomus temporalis*), entailed the strategic widening of roadside reserves. Monitoring of these management interventions, as well as monitoring of bird populations, highlighted the outstanding success of such efforts. Similarly, a carefully planned monitoring program within Booderee National Park on coastal New South Wales demonstrated the strong positive response of the highly endangered Eastern Bristlebird to a coordinated baiting program for the Red Fox (*Vulpes vulpes*) (Lindenmayer *et al.* 2009). These examples highlight the importance of monitoring in demonstrating management success.

2. It is now possible to identify some simple attributes characteristic of successful monitoring programs

Recently, Gene Likens and I (Lindenmayer and Likens 2010) completed a major review of monitoring and identified seven key characteristics of effective monitoring programs. These were: (1) Good questions and evolving questions. (2) The use of a conceptual model of an ecosystem or population. (3) Well-developed partnerships between scientists, policy makers and resource managers. (4) Strong and dedicated leadership. (5) Ongoing funding. (6) Frequent use of data. (7) Maintenance of data integrity and calibration of field techniques.

General recognition of these characteristics is important because it means that it is possible to identify ways to guide the design of new monitoring programs as well as examine and then improve the performance of existing monitoring programs. Moreover, the seven characteristics listed above are generic and are therefore relevant to any biodiversity monitoring program, irrespective of species, ecosystem or location.

3. There is an increased recognition among government and non-government agencies of the importance of monitoring

The importance of monitoring has long been recognised by the scientific fraternity (despite many scientists not engaging in monitoring programs – see Lesson 6 below). Although biodiversity monitoring has been acknowledged as important by policy makers and resource managers, this has often not converted to on-ground, high-quality monitoring programs. However, this has changed in many organisations in more recent times. For example, the Terrestrial Ecosystem Research Network (TERN), funded through the Australian Government Department of Innovation, Science and Research, commenced major funding in 2011 to maintain and/or expand long-term plot networks and environmental gradient transects in many parts of Australia. Formal environmental monitoring will be a significant component of this initiative.

In another example, in the Australian Federal Government's *Biodiversity Conservation Strategy*, it is stated that the nation will have a national long-term biodiversity monitoring and reporting system by 2015. Notably, the Federal Government has established formal monitoring programs as a fundamental component of several of its large-scale, long-term environmental initiatives. One of these is the Box-Gum Grassy Woodlands stewardship program which is paying private landholders to manage remnant areas of high quality woodland for conservation outcomes and in which a rigorously designed monitoring program will determine the effectiveness of management interventions (Zammit *et al.* 2010). Other organisations, like Catchment Management Authorities in NSW, are undertaking similar kinds of integrated management and monitoring initiatives. Of course, non-government organisations like Bush Heritage Australia and the Australian Wildlife Conservancy are similarly undertaking high quality management interventions accompanied by well-designed and implemented monitoring programs.

Monitoring failures

4. There are far too few well-designed and implemented long-term biodiversity monitoring studies in Australia

Despite the successes listed above, and the awakening recognition in some organisations of the need to undertake monitoring programs, the sad reality is that examples of effective biodiversity monitoring are far too rare in Australia. This problem is intimately linked with the fact that monitoring is almost invariably the last item funded in any environmental initiative; funding levels are never appropriate to do the monitoring well; funding is almost always too short term for work to be truly effective (e.g. a program exceeding 5–10 years; see Lindenmayer and Likens 2010), and funding for monitoring is invariably the first item cut from budgets. Our recent, questionnaire-based surveys of the people who are undertaking long-term ecological work in Australia clearly indicate that almost all projects are maintained and funded on an ad-hoc basis. These monitoring programs are highly susceptible to being terminated when the champion for the project leaves or dies, or when priorities change within the organisation with which the project champion is associated (Lindenmayer and Likens 2010).

The scarcity of good biodiversity monitoring programs means that it is impossible to report effectively on state and national environmental initiatives like the State of the Environment reports or to gauge the effectiveness of major tax-payer funded programs like the Natural Heritage Trust (Hajkowicz 2009).

5. There has been a demonstrable failure to slow biodiversity decline in Australia

Related to the previous point, and notwithstanding the successes of some management interventions for some species, there has been a demonstrable failure to slow the loss of biodiversity in large parts of Australia (Kingsford *et al.* 2009). The decline of native mammals in northern Australia is one of many sobering examples (Fitzsimons *et al.* 2010). This means that, despite the fact that over 10 per cent of the Australian land mass is in reserves, and the extent of protected areas has been increasing in the past few decades, many threatened species are not well protected in those reserves (Watson *et al.* 2010).

6. There has been a failure by a large proportion of the scientific fraternity to engage in monitoring

Successful monitoring typically involves partnerships between scientists, policy makers and resource managers (Russell-Smith *et al.* 2003; Lindenmayer and Likens 2010). Despite almost every ecologist and/or environmental scientist at some stage writing about the need to undertake monitoring of some form or another, many scientists have refused to actively engage in monitoring programs to guide: (1) their design; (2) the development of appropriate protocols; (3) the collection of data; or (4) data analysis and data interpretation. This is unfortunate as it has led to important monitoring programs either not being instigated, failing after they have commenced, or being poorly implemented.

Scientists also have often failed to highlight the dangers of **not** doing monitoring, such as the failure to detect important environmental changes (Lindenmayer *et al.* 2010) and the enormous logistical and financial difficulties associated with reversing some environmental problems (such as those associated with populations of invasive plants and animals) (McNeely *et al.* 2003; Simberloff 2010).

Connected with the lack of engagement of scientists in biodiversity monitoring has been a loss of new generations of people with the requisite natural history skills for effective empirically based biodiversity monitoring (Noss 1996; Lindenmayer and Likens 2010).

Solutions

7. New forms of long-term funding arrangements must be developed to support high quality, biodiversity monitoring programs in Australia

As outlined above, 'last thing funded/first thing cut' has been the overarching characteristic of the vast majority of monitoring programs around Australia. Monitoring programs do not function without appropriate levels of funding over an appropriate time frame. Short-term political initiatives and short-term funding cycles are therefore the antithesis of the kinds of funding streams and time frames needed to support and maintain effective ecological biodiversity monitoring programs (Lindenmayer 2007).

New kinds of long-term funding arrangements must be developed to support high-quality, biodiversity monitoring programs. The scale of action and level of these funding investments must match the seriousness of Australia's environmental and biodiversity conservation problems. Arguably, the plot and gradient-transect component of the national TERN initiative has made some tentative progress in this regard – but unfortunately it is only a 3-year funded program. As an example, I have recommended the establishment of a true environmental trust (that would support monitoring programs using the interest but not the capital of major investments) (Lindenmayer 2007). There are examples of environmental trusts and levies from around the world and there would be considerable value in 'cherry-picking' the best qualities of those that can be shown to have worked (Lindenmayer 2007). In addition, there also needs to be far more rigorous attention paid to ensuring that high-quality monitoring (by reputable individuals and organisations) is a mandatory part of any significant development proposal. Moreover, the level of investment in monitoring must be congruent with the scale of a development proposal. For example, a $50 000 biodiversity monitoring program in a multi-billion dollar natural resource extraction venture is an obscene mismatch.

Whatever the most appropriate and politically digestible funding formula ends up being, it is critical that the funding issues associated with establishing and then maintaining high-quality biodiversity monitoring should be resolved. Indeed, there will be few robust solutions to Australia's biodiversity problems (and indeed the nation's broader environmental problems) without tackling funding issues underpinning the maintenance of robust monitoring.

8. New institutional arrangements are urgently required to underpin effective biodiversity monitoring programs in Australia

The establishment and maintenance of effective biodiversity monitoring programs around Australia will require effective institutions and organisational structures. A new biodiversity monitoring-based institution is urgently required to fulfil a number of critically important roles. These include: (1) Mapping where and when past and existing monitoring efforts have taken place in Australia. (2) Coordinating and widely disseminating new ideas, protocols and methods about biodiversity monitoring, including providing opportunities for regular information exchange (e.g. conferences and workshops). (3) Providing advanced training in many of the key tasks associated with effective biodiversity monitoring, such as experimental design, application of field protocols, databasing, statistical analysis and interpretation, and scientific writing and reporting. This might include undergraduate training to provide high-quality professional staff to government and non-government agencies and corporate entities that will be undertaking biodiversity monitoring programs. (4) Lobbying for cultural changes within government bodies and other organisations so that they take biodiversity monitoring far more seriously and invest in it accordingly. And, (5) Assessing existing biodiversity monitoring programs to ensure they are improved to meet at least minimum quality standards (including proper and regular analysis and reporting of results).

9. New approaches are needed to link question-driven with mandated biodiversity monitoring programs

Recently, Gene Likens and I classified different kinds of monitoring (Lindenmayer and Likens 2010) and suggested that mandated monitoring occurs when environmental data are gathered as a stipulated requirement of government legislation or a political directive. Mandated monitoring is usually large scale (encompassing state or national level information) but does not attempt to identify or understand the mechanism influencing a change in an ecosystem or an entity. Rather, the focus is usually to identify trends (e.g. whether environmental conditions are getting 'better' or 'worse'). In contrast, question-driven monitoring programs often operate at the level of sites, landscapes or regions. These kinds of monitoring programs are guided by a conceptual model and by a rigorous experimental design. The use of a conceptual model will typically result in *a priori* predictions that then can be tested as part of the monitoring program. Often such learning is informed by strongly contrasting management interventions.

Different kinds of monitoring programs are conducted in different ways and at different spatial scales. A fundamentally important challenge remains about how to: (1) better integrate data, approaches and insights from different kinds of monitoring programs into useful environmental management, and, (2) use knowledge about the advantages and disadvantages of different kinds of monitoring programs to improve monitoring efforts.

I believe that many of the arguments in the scientific and resource management literatures about the suitability or otherwise of particular monitoring frameworks, as well as criticisms about particular monitoring programs being ineffective, stem from a failure to recognise the inherent values of, and differences between, large-scale mandated monitoring programs and smaller-scaled question-driven monitoring programs. I also suggest the next major challenge for long-term monitoring is to work out how to combine the data sets, results and outcomes that are conducted at different scales, in different ways and by different groups to produce integrated assessments useful to decision makers.

Concluding comments

High-quality biodiversity monitoring is fundamental to truly tackling Australia's biodiversity and environmental management problems. The appalling record on biodiversity monitoring in this nation needs to be urgently addressed. It is critical that governments in Australia are held accountable to pledges to implement a national long-term biodiversity monitoring and reporting system by 2015. Major changes in institutional arrangements and funding streams will be essential for achieving this stated goal and convincing a wide range of stakeholders that real gains are being made and that there are real biodiversity returns resulting from conservation investments.

Acknowledgements

The Thomas Foundation, and in particular Max Bourke, were the instigators and supporters of the meeting that led to this chapter and the push to see much better biodiversity monitoring take place in Australia. Many of the insights into monitoring over the past years and decades have resulted from long-term collaborative partnerships with expert statisticians, particularly Professor Ross Cunningham, Associate-Professor Jeff Wood and Professor Alan Welsh. The author is indebted to Professor Gene Likens for many of the topics that are touched upon in this chapter.

Biography

David Lindenmayer is Professor of Ecology and Conservation Science in the Fenner School of Environment and Society at The Australian National University. He has worked on an array of long-term ecological projects in south-eastern Australia over the past three decades, including several that classify as monitoring. He has written extensively on forest ecology and management, woodland biology, conservation biology, landscape ecology, wildlife conservation and management.

References

Allen RB (1993) An appraisal of monitoring studies in South Island tussock grasslands, New Zealand. *New Zealand Journal of Ecology* **17**, 61–63.

Fitzsimons J, Legge S, Traill B and Woinarski J (2010). *Into oblivion? The disappearing native mammals of northern Australia*. The Nature Conservancy, Melbourne.

Garnett S and Lindenmayer DB (2011). Conservation science must engender hope to succeed. *Trends in Ecology and Evolution* **26**, 59–60.

Hajkowicz S (2009). The evolution of Australia's natural resource management programs: Towards improved targeting and evaluation of investments. *Land Use Policy* **26**, 471–478.

Kingsford RT, Watson JEM, Lundquist CJ, Venter O, Hughes L, Johnston EL, Atherton J, Gawel M, Keith DA, Mackey BG, Morley C, Possingham HP, Raynor B, Recher HF and Wilson KA (2009). Major conservation policy issues for biodiversity in Oceania. *Conservation Biology* **23**, 834–840.

Lindenmayer DB (2007). *On Borrowed Time: Australia's Environmental Crisis and What We Must Do About It*. CSIRO Publishing and Penguin Publishing, Melbourne.

Lindenmayer DB and Likens GE (2010). *Effective Ecological Monitoring*. CSIRO Publishing, Melbourne.

Lindenmayer DB, Likens GE, Krebs CJ and Hobbs RJ (2010). Improved probability of detection of ecological 'surprises'. *Proceedings of the National Academy of Sciences of the USA* **107**, 21957–21962.

Lindenmayer DB, MacGregor C, Wood JT, Cunningham RB, Crane M, Michael D, Montague-Drake R, Brown D, Fortescue M, Dexter N, Hudson M and Gill AM (2009). What factors influence rapid post-fire site re-occupancy? A case study of the endangered Eastern Bristlebird in eastern Australia. *International Journal of Wildland Fire* **18**, 84–95.

McNeely JA, Neville LE and Rejmanek M (2003). When is eradication a sound investment? *Conservation in Practice* **4**, 30–41.

Norton DA (1996). Monitoring biodiversity in New Zealand's terrestrial ecosystems. In *Papers from a Seminar Series on Biodiversity*. November, Christchurch. (Eds B McFadgen and S Simpson) pp. 19–41. Department of Conservation, Wellington, New Zealand.

Noss R (1996). The naturalists are dying off. *Conservation Biology* **10**, 1–3.

Robinson D (2010). Working with landholders to protect woodland birds: an 18-year lesson from northern Victoria. In *Temperate Woodland Conservation and Management*. (Eds DB Lindenmayer, AF Bennett and RJ Hobbs) pp. 335–341. CSIRO Publishing, Melbourne.

Russell-Smith J, Whitehead PJ, Cook GD and Hoare JL (2003). Response of *Eucalyptus*-dominated savanna to frequent fires: lessons from Munmarlary 1973–1996. *Ecological Monographs* **73**, 349–375.

Simberloff D (2010). Invasive species. In *Conservation Biology for All*. (Eds NS Sodhi and PA Ehrlich) pp. 131–152. Oxford University Press, New York.

Watson JE, Evans MC, Carwardine J, Fuller RA, Joseph LN, Segan DB, Taylor MFJ, Fensham RJ and Possingham HP (2010). The capacity of Australia's protected-area system to represent threatened species. *Conservation Biology* **25**, 324–332.

Zammit C, Attwood S and Burns E (2010). Using markets for woodland conservation on private land: lessons from the policy-research interface. In *Temperate Woodland Conservation and Management*. (Eds DB Lindenmayer, AF Bennett and RJ Hobbs) pp. 297–307. CSIRO Publishing, Melbourne.

2 ACCOUNTABILITY: WE'RE AN INDULGENT AND MARGINAL PROFESSION IF WE CAN'T MEASURE THE EFFECTIVENESS OF INVESTMENT IN ENVIRONMENTAL MANAGEMENT

John C.Z. Woinarski

Lesson #1. Success is possible – a monitoring program in Kakadu demonstrates impacts of fire management.

Lesson #2. We cannot report reliably on biodiversity trends in Australia.

Lesson #3. In most cases, we cannot audit the effectiveness of investments in environmental management.

Lesson #4. Short-term cycles for government funding and academic attention spans conspire against long-term security for monitoring programs.

Lesson #5. Design is critical.

Lesson #6. Involve the managers throughout.

Lesson #7. Adopt business models for environmental management.

Lesson #8. Establish a national coordinating facility for biodiversity monitoring.

Lesson #9. Sequester long-term funding for foundation biodiversity monitoring programs.

Introduction

I like monitoring. I like its required discipline. I like the way it incrementally reveals a large picture, like an advent calendar. I like its added dimensionality: ecological research is often the search for, and interpretation of pattern, and monitoring adds a complex, piquant and intriguing dose of time to that patterning, and to the challenge of its decoding. I like the way it shows the ebb and flow of different species, each beating to its own rhythm. I like the way its design can produce definitive evidence about impacts and management interventions, or beyond design can reveal the workings of unpredicted events. I like the way it can be a relay event, with the baton of personal responsibility changing over time, the legacy growing like an aged wine.

Notwithstanding such attraction, collectively we've been particularly poor at biodiversity monitoring in Australia. I suspect that this is for two main reasons: (1) in some academic contexts, monitoring is perceived as pedestrian, tedious and technical, lacking the high science and swift incisiveness of analytic research; and (2) monitoring may take far more time than government funding and thesis cycles, and be an unsuitable vehicle for the delivery of academic outputs at preferred frequency. The consequential relative lack of long-term biodiversity monitoring in Australia is a major impediment to conservation. In a continent distinctive for its marked but irregular climatic variations, we have little evidence about the 'natural' beat of biodiversity responses, necessary to provide context for the interpretation of any change measured over a short term. We have little evidence to report on the biodiversity responses to ongoing climate change. We have little evidence about the longer-term consequences of land clearance and modification. We have little evidence to measure the effectiveness of our environmental management investment and practice. Among even the few long-term monitoring programs we have, there is little or no coordination or integration, and consequently we have few or no robust indicators of national biodiversity trends. This deficiency mutes our voice and severely limits our capacity to engage in national policy and agendas, to counterpoint environmental trends with the far more crisply defined economic and social indicators.

Note that I've deliberately omitted the word 'monitoring' from the title of this short piece. I suspect that part of the reason for the limited amount, and limited potency in policy setting, of biodiversity monitoring in Australia is due to misconception around the image of monitoring. It is too widely perceived as a peripheral and abstract activity, undertaken – if at all – at the end of a funding cycle. Rather, it should be a mechanism for understanding and demonstrating how ecological processes work, a core business accounting tool, and a central and essential component of our environmental management.

Lessons, failures and solutions

There are already excellent reviews of biodiversity monitoring, which include consideration of the critical components of successful programs (Lindenmayer and Likens 2010). The following points should be seen largely as embellishing or complementing such detailed accounts.

Successes

1. Success is possible – a monitoring program in Kakadu demonstrates impacts of fire management

Kakadu National Park is one of Australia's premier conservation reserves. A broad-scale, plot-based monitoring program for plants and terrestrial vertebrates, established in 1996, provides potent information about biodiversity trends in the park, particularly in relation to the effectiveness of fire management (Russell-Smith et al. 2009).

I qualify the assignation of 'success' to this program, on the basis that monitoring should be considered as simply an ingredient – if a critical one – in the package of environmental management. In the case of Kakadu, the monitoring program has provided the most unequivocal evidence for current decline in the native mammal fauna of northern Australia (Woinarski et al. 2010), but judgement of success in a monitoring program should extend beyond the clarity of trend assessment to encompass its capability to steer an improved management that results in delivering positive biodiversity outcomes. In this case, the conservation management package is evidently failing.

A strength and limitation of the Kakadu monitoring program is that it focuses particularly on one (key) management issue: fire. This design has resulted in the production of evidence that is definitive about the biodiversity consequences of different fire management actions

(Russell-Smith *et al.* 2009, 2010; Woinarski *et al.* 2010). However, the monitoring results also demonstrate that fire explains only part of the current mammal decline, and the current monitoring design cannot provide incisive assessment of the impacts of those other factors. This tension – between economic design with tight focus on the one hand and the capability to provide information on all significant management questions on the other – may be a common challenge for monitoring programs. In the case of the Kakadu monitoring program, the design may be more or less appropriate in that it provides answers to the primary question that it was established to address: how should fire be managed? However, serendipitously it also provided evidence that identifies another (perhaps more important) management challenge, broad-scale decline of native mammal species. It may now be appropriate to complement the existing monitoring with further more specifically designed monitoring components that target management for mammal decline.

Failures

2. We cannot report reliably on biodiversity trends in Australia

Ecology is a more grounded and rational science than economics. However, while we – and all other countries – have many robust and interpretable indices of economic and social prosperity, we have no sensible indices of biodiversity condition embedded into our national accounts.

This is a failing that has been repeatedly recognised (e.g. in State of the Environment reporting, see Beeton *et al.* 2006), notwithstanding some brave remedial attempts (such as in the Australian Bureau of Statistics *Measuring Australia's Progress* reporting). Lacking such national biodiversity monitoring data, environmental trends are marginalised or problematic to interpret, and are inequitably considered in decision-making against harder economic indices (Wentworth Group of Concerned Scientists 2008). For biodiversity, the most widely used trend measure in Australia is probably the change in the number of threatened species. However, in most cases (a noteworthy exception is the retrospective analysis of Garnett and Crowley 2008) this index is very inexact given lags and non-comprehensiveness in the assessment of the threatened status of species, and the lack of population monitoring for many or most threatened species in Australia.

A part of the problem is that biodiversity responsibilities in Australia are highly fragmented. Currently, there is no single body – or collection of state bodies – charged with the systematic collation and reporting of biodiversity trends across the nation. There are no certified monitoring program standards. This problem endures notwithstanding its long recognition, many fruitless proposals to address it and a recent proliferation of agencies ostensibly charged with national environmental reporting (Lindenmayer and Likens 2010).

3. In most cases, we cannot audit the effectiveness of investments in environmental management

Biodiversity conservation is a moral responsibility. But it is also a business, with investment, portfolios of responsibilities, investors (mostly taxpayers) and potentially competing firms. But as a business, it has a major failing: that it is currently almost impossible to demonstrate the effectiveness of its investments, or to provide evidence that can be used to make reliable choices between management options.

This is a pervasive problem, operating at all levels from the major national conservation programs (ANAO 2007, 2008) through to the business of managing individual properties (although with notable exceptions, particularly among non-government conservation landholders: for example Bush Heritage Australia 2007; Legge *et al.* 2011). It is a deficiency with multiple causes. A primary cause is the lack of adaptive management models, but disregard for monitoring is a major component of that cause. Characteristically, targets (if they are defined at all) don't relate specifically to biodiversity outcomes, there is little attention (or resourcing) given to designing programs that can measure progress towards targets, little reporting of such progress, little capability to amend management in response to the monitoring results and little responsibility taken, or censure given, if targets are not met.

If we were a business, we'd have a very poor credit rating. This makes it far less attractive for government – or others – to invest in biodiversity conservation management. But maybe this issue is more complex than business unreliability. Could it be too discomfiting if we had better monitoring programs, and higher profiling of their results? Such may reveal more starkly that our biodiversity management is failing, and that our economy is progressing at the expense of our biodiversity. Could it be that our society prefers a blurred ignorance of this situation rather than to be explicitly confronted with the environmental cost of our increasing prosperity?

4. Short-term cycles for government funding and academic attention spans conspire against long-term security for monitoring programs

I write as extensive areas of the continent are flooded, a conspicuous reminder that Australia is characterised by irregular climatic variation that drives major biodiversity response. Such potent variability means that it is difficult (or perhaps futile) to attempt to draw inference or meaning from ecological data collected over short time periods. Many critical environmental processes, and many management challenges, play out over periods of decades or centuries. Conservation management demands patience and long-term planning, and it may take many years for the benefits (or detrimental consequences) of our actions to be realised, and especially so if the environment is irregularly buffeted by severe climatic variation.

Monitoring is a discipline, and interruptions or short-changing in an established monitoring regime may significantly compromise the overall program and its ability to document and explain trends. Monitoring is a long-term commitment. For the most part, governments operate on funding cycles that mismatch environmental time; and/or they prioritise activities that produce a tangible short-term outcome, rather than establishing or maintaining reporting programs that may not provide outcomes for many years. For the most part, academic institutions are also welded into short-term funding and projects. While these characteristics conspire against the establishment of secure long-term monitoring programs, they can be overcome: many other comparable countries have enduring long-term ecological research and monitoring programs (Likens and Lindenmayer 2011).

Solutions

5. Design is critical

A biodiversity monitoring program should have a clear and worthwhile purpose, preferably to provide an answer to an explicit management question or to provide the evidence to demonstrate the effectiveness of a specific management action, or set of actions.

Around that purpose, the design of the monitoring program should be a straightforward but critical exercise based on statistical rigour balanced with purpose-fit practicality. Pragmatically, such design may need some Goldilocks realism. Many monitoring programs pay too little attention to sampling design and statistical power, and hence may have little or no capability to resolve the question that they were meant to address: these may be futile exercises. In other cases, statistically elegant designs may be so expensive and cumbersome that they may never be funded or may become an insuperable burden.

Design is critical for good monitoring, but should not constipate the process. For example, there has been longstanding if lukewarm interest by state and national conservation agencies in a coordinated biodiversity monitoring program for Australian rangelands. However, there has been no progress in such a program, other than a series of reviews and aggregations of suggested components (Whitehead *et al.* 2001; Smyth *et al.* 2003, 2004): that is, ecologists have collected the recipe ingredients, but haven't found the connections or mechanism (or the cook) to combine and serve them. This perhaps reflects a more generic disconnect in Australia at least between conservation scientists (operating largely in academic institutions) and individuals and agencies involved in environmental policy and management (Gibbons *et al.* 2008).

6. Involve the managers throughout

Adaptive management (Holling 1978) provides a framework for the counterpointing and interplay of management and monitoring, whereby monitoring provides an explicitly tailored assessment of management performance and a basis for continually refining such management. A key part of this interplay is the involvement of managers in monitoring.

For example, the fire-monitoring program in Kakadu National Park was designed collaboratively with the park managers, rangers and Aboriginal traditional owners; all these people helped undertake the field work involved in the monitoring and helped analyse and report on the results. This design allowed rangers to see the sometimes distant or indirect consequences of their actions in lighting, suppressing or not suppressing fires: to show clear cause and effect in their management responsibilities. This involvement provides some 'ownership' of the monitoring program and direct personal link to management outcomes. For the management agency, the results of the monitoring program provide a robust annual measure of environmental performance, and form the basis for continuing evaluation of fire management practice. In turn, the management staff involvement and provision of relevant agency performance indices help provide the foundation for the institutional support that is necessary to maintain resourcing for a long-term monitoring program.

7. Adopt business models for environmental management

Biodiversity monitoring will be a marginal and poorly resourced indulgence so long as there is little accountability in environmental management, and little explicit direct responsibility taken or given for environmental outcomes.

The issue here is not so much to focus on biodiversity monitoring *per se*, but more to instil accountable business practice into environmental management. Because it involves direct and significant economic consequences from biodiversity management, perhaps the best example of biodiversity monitoring in Australia is that associated with commercial fisheries, most notably the regular publication of fisheries status reports for targeted (and in some cases, by-catch) fish (and other marine) species (e.g. Wilson *et al.* 2009). In most cases, fishing regulation mandates the establishment of appropriate monitoring programs and that harvest levels and other management actions must be regularly reviewed with regard to the monitoring results. Such a model should be applicable to biodiversity more generally; although of course it may be an imperfect model in that it is noteworthy that many fisheries worldwide have collapsed, notwithstanding the implications long before evident in monitoring data.

8. Establish a national coordinating facility for biodiversity monitoring

There are relatively few biodiversity monitoring projects being undertaken in Australia, but these are being conducted mostly idiosyncratically by Commonwealth and state agencies, non-government conservation organisations, universities and other groups. There is little consistency in methodologies, analysis, reporting, or data storage, and little or no attempt to coordinate. If we are to report on the state of our environment at national, state or regional levels, then there is a clear need to attempt to better coordinate these existing scattered projects (and to develop additional far more strategic national programs), including to record standard operating procedures, to maintain meta-data, to interpret and display data, and to aggregate and disaggregate trends. It is striking and embarrassing that there is better coordination of biodiversity monitoring projects across countries (notably such as the Living Planet Index: WWF 2010) than there is across the states and territories of Australia.

One specific component of such a national coordination of biodiversity monitoring should be for threatened species. Currently, and appropriately, much attention and resourcing is invested in these species; however, it is almost impossible to get any coherent picture of the effectiveness of this attention.

9. Sequester long-term funding for foundation biodiversity monitoring programs

It is self-evident that (most) monitoring should be for the long term, and hence must have long-term funding security.

There are several models for such longevity. One is through perpetual endowment, widely used in philanthropic educational and arts funding, whereby a large initial bequest can support regular off-takes from interest to support monitoring activities. Another model is to use environmental offsets, particularly those for long-term resource projects. One recent notable Australian example involves the environmental offset developed for the Gorgon petro-chemical development on Barrow Island, Western Australia, which mandates a monitoring component for marine turtles to be conducted for the life of the project (about 60 years) (Office of the Appeals Convenor 2009). Another example is the need for long-term environmental monitoring now required to demonstrate outcomes from carbon investment projects, which may typically require explicit demonstration of 'permanency' of environmental benefits (e.g. CCBA 2008; DCCEE 2010).

Conclusions

There's much that we don't do well in managing our lands and seas. Biodiversity monitoring is one of the more acute of these failures. Of course, there are some things that we do well, and we should celebrate and learn from those. But it is difficult to ascribe success, and its causes, if we haven't measured and interpreted success through sound monitoring programs.

Australia's Biodiversity Conservation Strategy 2010–2030 explicitly recognises the importance of, and commits to, biodiversity monitoring as one of the key ingredients to biodiversity conservation. This represents an unprecedented opportunity: the accommodation within a high-profile policy setting of what many Australian ecologists and conservation managers have long sought. The challenge is now fairly simple: how do we do biodiversity monitoring effectively; how can we ensure its ongoing support; how can we most effectively ensure collaboration between agencies; how can we most effectively integrate reporting and communication; and how can we use monitoring as a mechanism to achieve improved conservation management? All of these tasks are achievable, as this book should make clear.

Acknowledgements

I'm grateful to Sarah Legge, Stephen Garnett, Alaric Fisher, Tony Griffiths, Jeremy Russell-Smith, Peter Harrison, Gordon Guymer, Bill Humphreys, Guy Fitzhardinge, and Max Bourke for discussions about monitoring; to David Lindenmayer for inviting this contribution; to Phil Gibbons, Alaric Fisher and David Lindenmayer for comments on a previous draft; and to the Department of Sustainability, Environment, Water, Population and Communities, the Tropical Savannas Cooperative Research Centre, and the National Environmental Research Program for supporting my involvement in monitoring, particularly the Kakadu monitoring program.

Biography

At the time of writing this chapter, **Dr John Woinarski** was Director of the Biodiversity Conservation Division of the Northern Territory Department of Natural Resources, Environment, the Arts and Sport. Subsequently, he has resigned from the Northern Territory environment department, and now lives on Christmas Island. He remains an Adjunct Professorial Fellow with the Research Institute for the Environment and Livelihoods, Charles Darwin University. He has been concerned with biodiversity conservation issues in northern Australia for 20 years. His interest in biodiversity monitoring is also piqued by membership of Australia's Threatened Species Scientific Committee, which has responsibilities for the fate of threatened species in Australia.

References

ANAO (2007). 'The conservation and protection of national threatened species and ecological communities'. Report No. 31, 2006–07, Performance Audit. On-line: <http://www.anao. gov.au/download.cfm?item_id=8B64F1991560A6E8AAB74 A9978383784&binary_id=9B0 B8E391560A6E8AAD7E4533DAEF3E9> [Accessed 10/02/2011] Australian National Audit Office, Canberra.

ANAO (2008). 'Regional delivery model for the Natural Heritage Trust and the National Action Plan for Salinity and Water Quality'. Report No. 21, 2007–08, Performance Audit. On-line: <http://www.anao.gov.au/download.cfm?item_id=CE21 C4471560A6E8AA2773F E12C20785&binary_id=F20BFC111560A6E8AA75D180CD56E65 2> [Accessed 10/02/2011] Australian National Audit Office, Canberra.

Beeton RJS, Buckley KI, Jones GJ, Morgan D, Reichelt RE and Trewin D (2006). 'Australia state of the environment'. Independent report to the Australian Government Minister for the Environment and Heritage. Department of the Environment and Heritage, Canberra. <http://www.environment.gov.au/soe/2006/publications/report/pubs/soe-2006-report.pdf>

Bush Heritage Australia (2007). 'Restoring land and wildlife, south-west Western Australia'. Bush Heritage Australia, Melbourne.

The Climate, Community & Biodiversity Alliance (CCBA) (2008). *Climate, Community & Biodiversity Project Design Standards*. 2nd edition. CCBA, Arlington, VA.

Department of Climate Change and Energy Efficiency (DCCEE) (2010). 'Design of the Carbon Farming Initiative'. Consultation paper. DCCEE, Canberra.

Garnett ST and Crowley GM (2008). The history of threatened birds in Australia and its offshore islands. In *Contributions to the History of Australasian Ornithology*. (Eds WE Davis Jr, HF Recher, WE Boles and JA Jackson) pp. 387–439. Nuttall Ornithological Club, Cambridge, MA.

Gibbons P, Zammit C, Youngentob K, Possingham HP, Lindenmayer DB, Bekessy S, Burgman M, Colyvan M, Considine M, Felton A, Hobbs RJ, Hurley K, McAlpine C, McCarthy MA, Moore J, Robinson D, Salt D and Wintle B (2008). Some practical suggestions for improving engagement between researchers and policy-makers in natural resource management. *Ecological Management & Restoration* **9**, 182–186.

Holling CS (Ed.) (1978). *Adaptive Environmental Assessment and Management*. John Wiley & Sons, New York.

Legge S, Kennedy MS, Lloyd R, Murphy SA and Fisher A (2011). Rapid recovery of mammal fauna in the central Kimberley, northern Australia, following the removal of introduced herbivores. *Austral Ecology* **36**, 791–799.

Likens GE and Lindenmayer DB (2011). A strategic plan for an Australian long-term environmental monitoring network. *Austral Ecology* **36**, 1–8.

Lindenmayer DB and Likens GE (2010). *Effective Ecological Monitoring*. CSIRO Publishing, Melbourne.

Office of the Appeals Convenor (2009). 'Statement that a proposal may be implemented (pursuant to the provisions of the *Environment Protection Act 1986*). Gorgon Gas development revised and expanded proposal Barrow Island nature reserve'. Statement no. 800. Office of the Appeals Convenor, Perth, WA.

Russell-Smith J, Edwards AC, Woinarski JCZ, McCartney J, Kerin S, Winderlich S, Murphy BP and Watt F (2009). Fire and biodiversity monitoring for conservation managers: a 10-year assessment of the 'Three Parks' (Kakadu, Litchfield and Nitmiluk) program. In *Culture, Ecology and Economy of Fire Management in Northern Australia: Rekindling the wurrk Tradition*. (Eds J Russell-Smith, PJ Whitehead and P Cooke) pp. 257–286. CSIRO Publishing, Melbourne.

Russell-Smith J, Price OF and Murphy B (2010). Managing the matrix: decadal responses of eucalypt-dominated savanna to ambient fire regimes. *Ecological Applications* **20**, 1615–1632.

Smyth A, James C and Whiteman, G (2003). *Biodiversity Monitoring in the Rangelands: A Way Forward*. Environment Australia and CSIRO, Canberra.

Smyth AK, Foulkes J and Watt A (2004). Biodiversity monitoring in Australia's rangelands: introduction. *Austral Ecology* **29**, 1–2.

Wentworth Group of Concerned Scientists (2008). 'Accounting for nature. A model for building the National Environmental Accounts of Australia'. Wentworth Group of Concerned Scientists, Sydney.

Whitehead P, Woinarski J, Fisher A, Fensham R and Beggs K (2001). 'Developing an analytical framework for monitoring biodiversity in Australia's rangelands'. Tropical Savannas Cooperative Research Centre, Darwin.

Wilson D, Curtotti R, Begg G and Phillips K (Eds) (2009). 'Fishery status reports 2008: status of fish stocks and fisheries managed by the Australian Government'. Bureau of Rural Sciences & Australian Bureau of Agricultural and Resource Economics, Canberra.

Woinarski JCZ, Armstrong M, Brennan K, Fisher A, Griffiths AD, Hill B, Milne DJ, Palmer C, Ward S, Watson M, Winderlich S and Young S (2010). Monitoring indicates rapid and severe decline of native small mammals in Kakadu National Park, northern Australia. *Wildlife Research* **37**, 116–126.

World Wildlife Fund (WWF) (2010). 'Living Planet Report: biodiversity, biocapacity and development'. WWF, Gland, Switzerland.

3 ECOINFORMATICS SOLUTIONS TO SUPPORT MONITORING FOR IMPROVED BIODIVERSITY CONSERVATION

Andre Zerger and Warwick McDonald

Lesson #1. The Landsat satellite monitoring program is an enduring monitoring program with extensive international uptake.

Lesson #2. Birds Australia Bird Atlas data provide an exemplar of how citizen science can build an ecological database.

Lesson #3. The National Land and Water Resources Audit highlights how national and jurisdictional collaboration is essential for enabling ongoing monitoring.

Lesson #4. Institutional leadership and coordination is critical to support improved environmental data discovery and access to support monitoring.

Lesson #5. Antiquated environmental data discovery tools limit our ability to gain awareness about existing data in increasingly data-rich environments.

Lesson #6. Biodiversity-related field work is uniquely time-consuming and improvements are required to support more efficient data acquisition.

Lesson #7. We require one organisation to lead and integrate environmental data discovery and access activities.

Lesson #8. Achieving a shift from simple data discovery to data 'interactivity' will improve our ability to utilise existing environmental data.

Lesson #9. Electronic field data acquisition in the ecological sciences provides a tractable solution to improve data acquisition efficiency and quality.

Introduction

Establishing a foundation for more effective environmental monitoring is topical given the Australian Government's recent investment in environmental monitoring-related activities. These include the National Plan for Environmental Information (NPEI) initiative and programs supported by the National Collaborative Research Infrastructure Strategy (NCRIS) such as the Terrestrial Ecosystem Research Network (TERN), Integrated Marine Observing System (IMOS) and the Atlas of Living Australia (ALA). Pre-existing national-scale monitoring programs should not be overlooked including the Bureau of Meteorology's weather, climate and water monitoring programs, the Australian National Carbon Accounting System (NCAS) and the efforts led by the Australian Bureau of Statistics (national census, integrated economic-environmental accounts). At a global scale, initiatives such as the Group on Earth Observations Biodiversity Observation Network (GEO BON), DataONE, The Knowledge Network for Biocomplexity and the Global Index of Vegetation Plot Databases are developing strategies and tools to facilitate improved environmental data sharing and monitoring.

Robust discussion of environmental monitoring requires a taxonomy of monitoring types. There is a distinction between three types of monitoring, given below, that show that objectives differ and the relative weight placed on the components of a monitoring program can vary. The types of monitoring are not mutually exclusive but serve as generalisations to contextualise the discussion. These include the following:

- **Research monitoring** – Targeted research activities that create new empirical under-standing or validation of existing theory. This type of monitoring may target individual species or specific ecosystems and typically requires targeted experimental designs to address research questions. Although specific in aims, findings may be applied more widely as general biodiversity planning principles.
- **Program monitoring** – Monitoring activities designed to evaluate the effectiveness of resource management programs. They are tightly coupled with a policy imperative. For example, the Australian Government's National Carbon Accounting System (NCAS) is designed to serve the government's reporting requirements under the Kyoto Protocol and the UN's Framework Convention on Climate Change National Greenhouse Gas Inventories. Examples at a state level would include air quality monitoring (EPA Victoria's air quality monitoring program) or statewide land cover monitoring programs (e.g. Queensland's Statewide Land Cover and Trees Study – SLATS).
- **Foundation monitoring** – Monitoring de-coupled from any immediate research or program imperative that is established to provide a legacy of foundation data to support many needs (including research and program monitoring). National examples include the Birds Australia Bird Atlas project, the ALA and the NCRIS facilities named earlier. International examples include the suite of satellite-based imaging sensors deployed primarily by the US government, and providing freely accessible imaging products (e.g. Landsat, Advanced Very High Resolution Radiometer (AVHRR), Moderate Resolution Imaging Spectroradiometer (MODIS)).

The monitoring types help inform each other. Foundation monitoring can inform research monitoring how to best mobilise the community to complement its field surveys (e.g. Birds Australia Bird Atlas). Conversely, research monitoring provides insights regarding experimental design and stratification that can be key for establishing foundation monitoring programs such as TERN's plot network. Ultimately each approach to monitoring requires key decisions to be made regarding how to best invest limited resources into the elements of a monitoring infrastructure. Elements may include experimental design, field data acquisition (plot size, thematic detail, spatial replication, temporal replication), analysis, reporting, data documentation and dissemination and communication activities.

Ecoinformatics is defined in its broadest sense to include the application of computer technology to improve our ability to analyse and synthesise ecological data (Hale and Hollister 2009). These include the technical, such as: metadata, data standards, tools for discovery and access, and the institutional, including: licensing, governance, data custodianship responsibilities and institutional coordination. Our ability to monitor effectively is limited by lack of awareness and access to existing data rather than exclusively a need for new data. Therefore

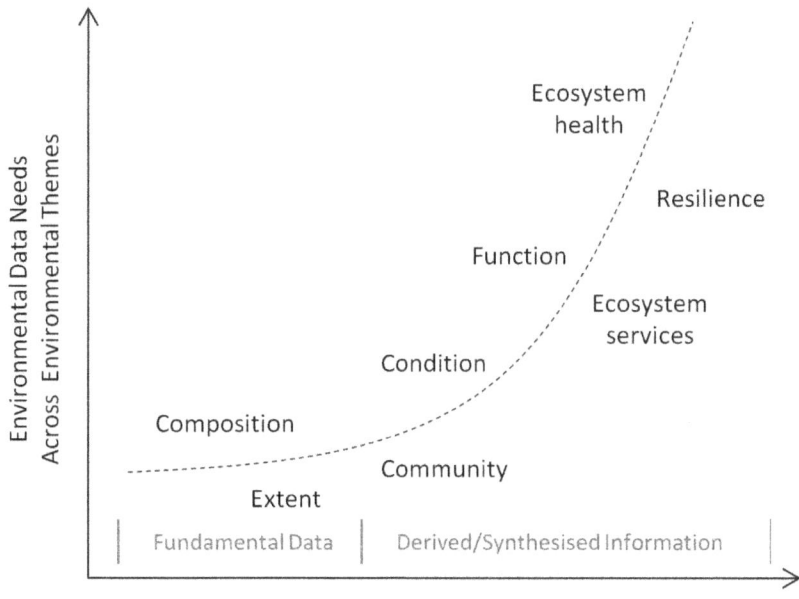

Figure 3.1. Relationship between higher levels of environmental system abstraction and data requirements across multiple environmental themes (e.g. air, land, water, agricultural production).

biodiversity monitoring, particularly at continental scales, can achieve improved and enduring effectiveness by placing a greater emphasis on the ecoinformatics elements of a monitoring system to overcome these limitations.

Madin *et al.* (2007) argue that ecological research (and this could be extended to monitoring) increasingly relies on aggregating data from existing studies to generate synthesised knowledge, and this integration demands ecoinformatics solutions. Initiatives such as the Collaboration for Environmental Evidence (http://www.environmentalevidence.org) and its focus on meta-analysis through systematic reviews highlight the importance of rapid access to existing data and knowledge. Figure 3.1 provides a schematic highlighting the relationship between environmental data needs and our ability to explore higher levels of environmental abstraction (beyond composition and extent) such as health, function, ecosystem services and resilience (Gibbons and Freudenberger 2006), or to support new monitoring frameworks such as integrated environmental–economic accounts, or the Wentworth Group's Accounting for Nature framework (Wentworth Group of Concerned Scientists 2008). Monitoring at higher orders of abstraction, particularly at continental scales can be partially addressed through ecoinformatics-led solutions. It is important to recognise that to support research monitoring, it may not be feasible to resource a detailed investment in informatics and other strategies may be required. Conversely, the existence of some key foundation monitoring programs which 'harvest' federated data is owed extensively to advances in ecoinformatics (e.g. ALA, GEO BON), and hence it is a proven enabler of monitoring.

The following discussion examines three successful monitoring programs and then presents some deficiencies, both technical and institutional, in how monitoring is supported for improved conservation outcomes. Finally, ecoinformatics-related solutions are offered as possible strategies for achieving improved monitoring.

Lessons

1. The Landsat satellite monitoring program is an enduring monitoring program with extensive international uptake

The NASA-led Landsat satellite program provides an exemplar of a highly effective foundation environmental monitoring program. Through its provision of reliable and quality-assured synoptic remote sensing data, its use for a variety of biodiversity applications – particularly at regional and continental scales – continues to be widespread nationally and internationally, even with the advent of newer satellite platforms. Its success can be attributed to its historical availability and reliability, albeit with some periodic quality or availability issues, and more recently, the fact that it has become freely available via the USGS GloVis service (http://glovis. usgs.gov). In Australia, Landsat forms the core of a number of key monitoring programs including the National Carbon Accounting System (NCAS) and Queensland's SLATS.

First, Landsat is a well-funded international monitoring program supported by a philosophy that supports free public access to data. So at one level, lessons from Landsat for monitoring biodiversity at regional scales may appear to offer little. However, putting aside the scale of the initiative and the philosophy of free access as an enabler of new approaches, Landsat provides other qualities of value to biodiversity monitoring. Users have access to primary (non-aggregated) sensed data and sensor parameters. This means new algorithms can extract further information from old imagery to gain insights about change across a broader historical context. This is an important lesson for biodiversity monitoring as it highlights the value of preserving primary data, and ensuring methodologies for collection are documented and accessible. Second, although Landsat may be limited in its spatial resolution for some applications, the temporal repeat can partially offset this. For example, in the case of vegetation mapping, knowledge of species phenology and access to time series, temporal Landsat can be exploited to improve the accuracy of mapping.

2. Birds Australia Bird Atlas data provides an exemplar of how citizen science can build an ecological database

There are analogues between foundation monitoring programs such as the Birds Australia Bird Atlas project (Barrett *et al.* 2003) and initiatives such as Kingsford's waterbird surveys (Kingsford 1999) and thematically unrelated activities such as the Australian Bureau of Statistics' National Census. First, they highlight that although their thematic detail may be limited, particularly when compared to typical research monitoring programs, their spatial extent makes them valuable and they rapidly become central to supporting research monitoring. Second, as in the case of the Landsat program, the historical time series they eventually generate provides an essential resource for a suite of monitoring needs. The Birds Australia program highlights the value of investing in foundation continental-scale monitoring without necessarily requiring an exhaustive inventory of possible uses, or business drivers. Evidence shows that continental-scale enduring foundation monitoring programs, albeit limited in their thematic detail, become increasingly invaluable monitoring resources.

3. The National Land and Water Resources Audit highlights how national and jurisdictional collaboration is essential for enabling ongoing monitoring

The Australian National Land and Water Resources Audit (NLWRA) was established in 1997 and sponsored by the Natural Heritage Trust to 'provide the baseline for the purposes of carrying out assessments of the effectiveness of land and water degradation policies and programs … and … to improve Commonwealth, State and regional decision-making on natural resource management'. The NLWRA was designed to address both foundation and program monitoring imperatives. It established some enduring legacy monitoring programs across a variety of environmental themes that still remain active, including the Australian Soil Resources Information System and the National Vegetation Information System. Some, such as the Australian Collaborative Rangelands System, have provided an enduring legacy of time series biophysical data. The success of these initiatives, particularly in terms of their longevity and fostering of national standards for environmental data collation, can be attributed to a successful national cooperative approach, with strong jurisdictional partnerships and ongoing engagement with the science community. Through these pathways the NLWRA made significant progress in institutionalising monitoring, evaluation and reporting.

4. Institutional leadership and coordination is critical to support improved environmental data discovery and access to support monitoring

Reichman and Jones (2011) argue that advances in ecoinformatics will transform ecology by facilitating 'cross-cutting' analysis and synthesis of existing data. Cross-cutting analysis across multiple environmental themes (e.g. air, land, oceans and water) is essential to monitor change at higher levels of environmental abstraction (see Figure 3.1). Rapid access to the best available environmental data is key to conducting continental-scale monitoring which relies on existing data, given the time and resources required to implement new programs at such scales. This challenge can be felt more acutely by staff in organisations relatively removed from Australian Government and research monitoring activities (program and research monitoring). However, even within the Australian Government and the research community, the creation of new ecological monitoring programs, increasing prevalence of environmental data produced through environmental modelling, in-situ sensor networks and high-temporal resolution synoptic remote sensing are rapidly creating new environmental data discovery challenges. There are technical challenges with improving users' awareness of, and access to, existing environmental data. However, there is a pressing need for institutional leadership and coordination to establish the organisational frameworks to allow users to harness our existing environmental data holdings.

5. Antiquated environmental data discovery tools limit our ability to gain awareness about existing data in increasingly data-rich environments

Environmental data discovery catalogues do exist in some themes such as oceans data, via the Australian Ocean Data Network (http://www.aodc.gov.au), or AuScope (http://www.auscope.org.au) for earth-sciences data. Some Australian Government agencies with a data custodianship and operational data management responsibility also maintain operational systems (e.g. the Bureau of Meteorology and Geoscience Australia). However, for some environmental themes such as land (terrestrial ecosystems including soils, vegetation and fauna) there is a general absence of functional catalogues and users must utilise manual discovery methods. Deficiencies in data access and discovery have also been recognised at the whole of Australian Government level through its Gov 2.0 initiative (http://data.gov.au). Tools such as the Australian Spatial Data Directory (ASDD) can partially address this. However, it is difficult to know whether the ASDD provides an exhaustive and up-to date list of data across environmental themes or regions or if it has a major focus on only spatial data, and not all agencies publish or update to the ASDD.

Existing tools that support environmental data discovery (e.g. catalogues and portals) remain relatively simplistic, demanding an *a priori* understanding of the data the user is seeking (e.g. title, custodian and geographic location). This traditionally involves a one-off search by prospective users with limited ongoing interaction between the user, the data and producer. This model of data discovery served the community relatively well in an era of relative static data upgrades and only intermittent changes in data availability. However, in an era of rapid data upgrades, the establishment of new monitoring programs (e.g. TERN) and high temporal frequency satellite imaging, remaining cognisant of changes in data is challenging in data-rich environments. There is consequently a need for more intelligent data interactivity, rather than simple discovery tools.

6. Biodiversity-related field work is uniquely time-consuming and improvements are required to support more efficient data acquisition

Monitoring to support biodiversity conservation introduces challenges not readily addressed by leveraging lessons from related environmental monitoring programs, or aggregating knowledge about existing data holdings into centralised data discovery and access tools. Monitoring the abundance of fauna at continental scales is thematically more complex than monitoring attributes of a hydrologic system with stream gauges. Creating a vegetation map from sparse field observations to generate regional or continental-scale surfaces, as is done for rainfall, is not possible when human modification continues to alter vegetation cover and distribution. Biodiversity monitoring introduces unique challenges, foremost of which is the time-consuming nature of biodiversity field work in increasingly resource-constrained settings. There is a need to achieve improvements in the way we acquire biodiversity data to support monitoring programs. Achieving improvements in the efficiency of primary data acquisition can enable greater spatial and temporal replication, improved data accuracies and an ability to more rapidly progress from data acquisition to analysis, reporting and decision-making.

7. We require one organisation to lead and integrate environmental data discovery and access activities

Developing solutions to coordinate knowledge about existing environmental data holdings across Australia is initially an institutional challenge. Challenges include (1) agreement on common standards, (2) universal approaches to data licensing, (3) common models for structuring information, (4) setting data priorities, (5) establishing a commitment to resourcing a shared model for national data discovery, and (6) governance arrangements to define roles and responsibilities. One solution is to establish an organisation to provide this ongoing leadership role, initially with a focus on Australian Government-generated environmental data, but the scope could be broadened to environmental data generated by a variety of Australian Government programs (e.g. Caring for our Country, Australian Research Council). Although ambitious in scope, there is precedent for this within environmental themes including the Atlas of Living Australia (non-ongoing), the Bureau of Meteorology's ongoing responsibilities with regard to climate and water, and the Australian Bureau of Statistics. Internationally, there is evidence that organisations are moving to improved models of coordination and data management for government research and related data, for example the INSPIRE Directive (http://inspire.jrc.europa.eu) in Europe and the US National Science Foundation (Nelson 2009).

8. Achieving a shift from simple data discovery to data 'interactivity' will improve our ability to utilise existing environmental data

The concept of environmental data 'interactivity' seeks to develop a more intelligent and enduring relationship between the user, data and the data producer to support more effective data use in increasingly data-rich environments. There are institutional and technical elements to this challenge. If institutional arrangements are in place, there are opportunities to innovate how one supports user interactivity and data discovery with next generation catalogues. The proliferation of internet-enabled intelligent search engines and associated indexing engines (e.g. Google); e-commerce facilities supported by 'recommendation engines' (e.g. Amazon's 'people who bought this also bought ...' feature); exploiting the thinking and design principles emerging from social networking technologies to support active communities of practice around key environmental data and specifically for those national data sets which are resourced for regular updates and improvements; or utilising technologies employed by journal publishers such as citation indexes and cross-citation metrics in an environmental data-use setting, all these can provide useful analogues for how to practically design systems to shift from a static model of data discovery to one of interactivity through the entire environmental data life cycle.

9. Electronic field data acquisition in the ecological sciences provides a tractable solution to improve data acquisition efficiency and quality

Preceding discussions have highlighted the uniqueness of biodiversity data in its thematic detail and the consequent expense of ecological data acquisition for all types of monitoring. This can be overcome through technologies such as wireless environmental sensor networks (Collins *et al.* 2006). However, these technologies remain so immature in ecological domains that a long journey awaits their use (Langendoen *et al.* 2006; Zerger *et al.* 2010). Alternatively, community-resourced monitoring programs such as the Birds Australia Atlas project may provide opportunities for efficiencies. However, birds engender a level of community support for monitoring that would be impossible to replicate for other species. Similarly, less structured 'crowdsourcing' solutions (Fritz *et al.* 2009) are yet to prove their value, let alone for biodiversity monitoring.

Gains towards improved monitoring can be achieved by developing and applying field tools that enable experts (botanists, zoologists) to electronically capture field data through the use of tablet and handheld computers coupled via data-entry tools in real-time to centralised databases. Electronic data acquisition minimises transcription errors and constructs a database in real-time, enabling almost immediate analyses and quality assurance as the monitoring program evolves. It also removes the need for paper transcription and it can seamlessly make primary ecological data instantly available to a broader monitoring community. The prevalence of tablet computers and the increasing availability of wireless internet access in rural areas is making this feasible. Digital provision of ecological data from the field to the laboratory provides numerous benefits. However, the primary barrier to the use of such tools is the potential users' ability to develop and deploy a customised electronic field-data acquisition system (data capture forms and relational database technologies). Investing in developing such systems for the ecological sciences, while borrowing on advances in other areas is critical to improve our ability to monitor rapidly and accurately, and to facilitate rapid data access. Where appropriate, the logical extension of these technologies is to support 'citizen science' initiatives as a pathway for constructing and updating ecological data sets.

Conclusion

In this chapter we have focused on ecoinformatics-based solutions and related institutional challenges to achieving improvements in biodiversity monitoring in Australia, with the discussion framed in a relatively resource-unconstrained context. Aspirational goals have been intentionally set for improving our ability to monitor through ecoinformatics-led transformations recognising that pursuing these solutions will be challenging for some monitoring types. If evidence from other domains such as health, finance and utilities management serves us well, a shift towards improved coordination and leadership to support more intelligent data discovery and access will eventually occur in the environmental sciences. Our challenge is to ensure these transformations become priorities and that they are designed in ways which practically support our biodiversity monitoring needs, whether framed in the context of species composition, extent, condition, ecosystem services or abstractions yet to be developed.

Acknowledgements

Thanks are extended to Jon Millard and Rob Vertessy for their insights that helped shape the paper. Natasha Herron provided reviews of early drafts of the manuscript. We are grateful to David Lindenmayer and Philip Gibbons for providing us an opportunity to share our thoughts and vision.

Biographies

Andre Zerger is Section Head, Environmental Information Services with the Bureau of Meteorology. His interests focus on the development of spatial analysis and remote sensing methods to support natural resource management assessment and monitoring, with a specific focus on biodiversity and conservation planning (particularly native vegetation condition monitoring). He is leading activities under the Bureau of Meteorology's contribution to the National Plan for Environmental Information initiative.

Warwick McDonald is Branch Head, Environmental Information Services within the Bureau of Meteorology. He recently headed the Water Information Research and Development Alliance (WIRADA) between the Bureau of Meteorology and the Commonwealth Scientific and Industrial Research Organisation (CSIRO). Warwick leads the Bureau of Meteorology's contribution to the Australian Government's National Plan for Environmental Information initiative.

References

Barrett G, Silcocks A, Barry S, Cunningham R and Poulter R (2003). *The New Atlas of Australian Birds*. Royal Australasian Ornithologists Union, Melbourne.

Collins SL, Bettencourt LMA, Hagberg A, Brown RF, Moore DI, Bonito G, Delin KA, Jackson SP, Johnson DW, Burleigh SC, Woodrow RR and McAuley JM (2006). New opportunities in ecological sensing using wireless sensor networks. *Frontiers in Ecology and the Environment* **4**, 402–407.

Fritz S, McCallum I, Schill C, Perger C, Grillmayer R, Acchard F, Kraxner F and Obersteiner M (2009). Geo-Wiki.org: The use of crowdsourcing to improve global land cover. *Remote Sensing* **1**, 345–354.

Gibbons P and Freudenberger D (2006). An overview of methods used to assess vegetation condition at the scale of the site. *Ecological Management & Restoration* **7 S1**, 10–17.

Hale SS and Hollister JW (2009). Beyond data management: how ecoinformatics can benefit environmental monitoring programs. *Environmental Monitoring and Assessment* **150**, 227–235.

Kingsford RT (1999). Aerial surveys as a measure of river and landscape and floodplain health. *Freshwater Biology* **41**, 425–438.

Langendoen KG, Baggio A and Visser OW (2006). Murphy loves potatoes: experiences from a pilot sensor network deployment in precision agriculture. *20th International Parallel and*

Distributed Processing Symposium Proceedings. 25–29 April 2006, Rhodes Island, Greece. IEEE, New England, USA.

Madin J, Bowers S, Schildhauer M, Krivov S, Pennington D and Villa F (2007). An ontology for describing and synthesizing ecological observation data. *Ecological Informatics* **2**, 279–296.

Nelson B (2009). Data sharing: Empty archives. *Nature* **461**, 160–163.

Reichman OJ and Jones MB (2011). Challenges and opportunities of open data in ecology. *Science* **331**, 703–705.

Wentworth Group of Concerned Scientists (2008). 'Accounting for nature. A model for building the National Environmental Accounts of Australia'. Wentworth Group of Concerned Scientists, Sydney.

Zerger A, Viscarra-Rossell R, Swain D, Wark T, Handcock R, Doerr VJ, Bishop-Hurley G, Doerr E, Gibbons P and Lobsey C (2010) Proximal environmental sensing for vegetation, animal and soil sciences. *International Journal of Applied Earth Observation and Geoinformation* **12**, 303–316.

4 MONITORING AUSTRALIAN BIRDS TO MEET INTERNATIONAL OBLIGATIONS

Stephen T. Garnett

Lesson #1. Volunteers have been instrumental in bird monitoring in Australia, producing results that have had substantial impact on policy and practice both nationally and internationally.

Lesson #2. Empowerment of local communities can lead more readily to changes on the ground.

Lesson #3. Efficient data management and timely analysis and reporting are critical to effective action.

Lesson #4. Skills in interpreting monitoring results are required in government as well as among volunteers and researchers.

Lesson #5. Australia will be able to advocate monitoring internationally more effectively if it does more itself.

Lesson #6. Develop metrics to measure skill levels needed by government and those they currently have.

Lesson #7. Develop balanced models of government–NGO monitoring partnerships.

Lesson #8. Use some monitoring resources employing people to read and organise published literature.

Introduction

Australia is party to 33 international agreements and conventions that aim to promote the protection of biodiversity (Natural Resource Management Ministerial Council 2010). Every one of these agreements requires monitoring to assess compliance and performance, but for few is there substantial or systematic government investment. However, some monitoring by volunteer groups has been particularly comprehensive and some new government programs have been initiated recently. Here, I concentrate on Australia's performance on monitoring of agreements that relate specifically to birds, of which there are eight (see Table 4.1).

The most inclusive of these agreements is the Convention on Biological Diversity (CBD) that aims to achieve a significant reduction in the current rate of biodiversity loss. This objective has also been adopted as a target under the Millennium Development Goals. Key means of reporting against these goals adopted by the CBD are the International Union for

Table 4.1. International agreements, conventions and partnerships relating to birds to which the Australian Government is signatory and organisations that undertake monitoring for each of these

Date of agreement	Name	Monitoring of birds relevant to agreement
1971	Convention on Wetlands of International Importance especially as Waterfowl Habitat (Ramsar Convention)	Australian Wader Study Group and regional groups, Birds Australia
1979	Convention on the Conservation of Migratory Species of Wild Animals (CMS/Bonn Convention)	Australian Wader Study Group and regional groups, Birds Australia
1981	Japan–Australia Migratory Bird Agreement (JAMBA)	Australian Wader Study Group and regional groups, Birds Australia, Australian Government
1988	China–Australia Migratory Bird Agreement (CAMBA)	Australian Wader Study Group and regional groups, Birds Australia, Australian Government
1993	Convention on Biological Diversity (CBD)	Australian Wader Study Group and regional groups, Birds Australia
1997	International Plan of Action for Reducing Incidental Catch of Seabirds in Longline Fisheries	Australian Government, Tasmanian Government
2001	Agreement on the Conservation of Albatrosses and Petrels	Australian Government, Tasmanian Government
2006	The Partnership for the Conservation of Migratory Waterbirds and the Sustainable Use of their Habitats in the East Asian–Australasian Flyway (Flyway Partnership)	Australian Wader Study Group and regional groups, Birds Australia, Australian Government
2007	Republic of Korea–Australia Migratory Bird Agreement (ROKAMBA)	Australian Wader Study Group and regional groups, Birds Australia, Australian Government

the Conservation of Nature (IUCN) Red List Index (United Nations 2011) and the proportion of Important Bird Areas (IBAs) that are protected (United Nations 2010). Australia also has a range of agreements that aim to protect two groups of migratory birds. Bilateral agreements with Japan, China and South Korea, a multilateral involvement with the East Asian–Australasian Migratory pathway and the broader Ramsar Convention on Wetlands all have connections to a group of about 30 migratory shorebird species that regularly fly between the northern Palaearctic and Australian wetlands, particularly coastal mudflats. The monitoring for birds under these agreements has historically fallen to volunteers associated with Birds Australia and its subsidiary groups. A second group of agreements is primarily concerned, in Australia, with albatrosses and large petrels including specific agreements to conserve them and reduce incidental by-catch from fishing and the wider Bonn Convention on Migratory Species. Here the monitoring has primarily been conducted by government. Experience with monitoring birds for these conventions and agreements has relevance for other biodiversity classes.

Successes

1. Volunteers have been instrumental in bird monitoring in Australia, producing results that have had substantial impact on policy and practice both nationally and internationally

When in-kind and cash support for monitoring are combined for threatened birds in Australia, the volunteer contribution is at least as much as government (Garnett *et al.* 2003). For migratory shorebirds in particular, which were not threatened but are now, the ratio between volunteer and government support is at least ten to one (Garnett *et al.* 2011). Shorebirds have been monitored in Australia with increasing sophistication since the 1970s (Geering *et al.* 2007) and the information gathered has been instrumental to the protection of important sites for these birds around Australia as well as highlighting the critical stopover sites along migratory pathways.

Starting with a few people from Great Britain with skills in catching and marking migratory waders, the interest in this specialised group of birds has burgeoned not just around Australia but, as a result of training and collaboration, along the entire East Asian–Australasian migration route. Monthly surveys of sites near settlements and regular expeditions to remote parts of Australia, coupled with increasingly innovative use of technology, have delivered a clear picture of the location, size and temporal importance of sites for most of the migratory shorebirds on the flyway. These data have underpinned the definition of many Ramsar sites and are now used in international advocacy with China and South Korea to try to save critical habitat. While government often supported this work, the drive and energy came from the dedicated volunteers who collected and collated the data and are reporting on the information in increasingly high impact journals.

2. Empowerment of local communities can lead more readily to changes on the ground

There are now many examples showing that if those who have direct management responsibility for an area are also involved in research, they are more likely to take action on any findings arising from that research (e.g. Garnett *et al.* 2009). This is the principle behind the monitoring framework developed by BirdLife International for Important Bird Areas. BirdLife has adopted a system that can be deployed by people with relatively low skill levels to assess trends in the condition of IBAs, the threats they face, and the actions being taken to conserve them (State–Pressure–Response) on a scale of 0–3. Such scores are often composites: e.g. response measures are the total scores for the levels of formal protection, management planning and implementation of conservation action. As an example, a recent review of Kenyan IBAs over six years estimated that the average condition declined, threats eased slightly and responses increased (Mwangi *et al.* 2010). In this instance, local people made initial assessments of each IBA that were then collated and reviewed by a national coordinator and a national assessment team. Such monitoring is now being instituted globally; Birds Australia employs a national IBA Coordinator to aggregate results for Australia for the 314 IBAs recently identified for Australia and its offshore territories (Dutson *et al.* 2009).

IBA monitoring serves two critical functions. The first is that crude measures of local trends become more powerful when aggregated so that national or regional trends in biodiversity, threats or conservation response become statistically significant because of the number of

IBAs involved. Second, wherever possible, those monitoring IBAs are people living near the IBA and thus often a part of the communities generating some of the threats. By witnessing changes in IBA conditions themselves, people involved in the monitoring can initiate local debate in a way that would not happen were outsiders to undertake all the monitoring themselves (e.g. Burger and Liner 2005). IBAs are already starting to be used in advocacy in Australia by local groups, including Indigenous groups wishing to emphasise the value of their land as Indigenous Protected Areas. In terms of the trade-off between skill/sophistication and local ownership, BirdLife have opted for local ownership.

3. Efficient data management and timely analysis and reporting are critical to effective action

One of the characteristics of the Australasian Wader Studies Group (AWSG) and its subsidiaries is data-rich reporting. Every issue of the Bulletin of the AWSG, *Stilt*, has articles that provide detailed counts at different sites, recapture data, or maps. This has been possible because data management and reporting were given a priority by wader enthusiasts from the start. The *Stilt* is also an effective record of progress with wader studies in the region, and became a peer-reviewed journal with its 50th issue in 2008 representing the interests of the whole East Asian Australasian Flyway, not just Australasia. A newsletter, *The Tattler*, now carries much of the news that was in *Stilt* and is available on the web. The advantage of these communication organs is not just the information that they carry, but the feedback they provide to information providers and users. The capacity of wader study groups to produce reports rapidly and in a timely manner has undoubtedly contributed to ongoing volunteer enthusiasm – an enthusiasm that is remarkable given these birds are mostly brown, difficult to identify and live in habitats that are often uncomfortable and physically demanding to survey. Any national biodiversity monitoring strategy will need to provide the same richness in feedback to contributors if it is to maintain popular, and political, support.

Deficiencies

4. Skills in interpreting monitoring results are required in government as well as among volunteers and researchers

For many decades, the Commonwealth Scientific and Industrial Research Organisation (CSIRO) and all state and territory governments employed researchers to study wildlife. These government officers were then available to provide advice to government on wildlife management and biodiversity conservation based on personal experience, interest and professional training. Unfortunately, this skill base is gradually ageing and retiring, and is not being replaced with people from the same background.

The status of Australian birds has been assessed in 1990, 2000 and 2010 in collaboration with government, researchers and community groups (Garnett 1992; Garnett and Crowley 2000; Garnett et al. 2011). In five of the seven Australian states and territories, the same people who were involved in helping assess the status of threatened birds in 1990 have been closely involved two decades later, although they have changed positions within their institutional hierarchies. Sadly, none of the departments to which they belong are hiring new staff with similar skills and responsibilities. In the other two jurisdictions, there has been frequent turnover during this 20-year period and there is little institutional memory or appreciation of the assessment process among current staff. Overall there is a tendency for technical staff with experience in biodiversity conservation to be replaced with multi-skilled managers who can shift responsibilities easily in response to changes in government policy. As a result of these trends in employment policy, many government officers seem not to appreciate what information they need or, if information is obtained through contract, appreciate the significance or reliability of the findings. Given that conservation managers draw strongly on personal experience to make decisions (Pullin and Knight 2005), poor outcomes are inevitable if people without experience in conservation management are given responsibility for making decisions about whether monitoring should be funded, what to monitor and how the results should be interpreted and incorporated into policy.

5. Australia will be able to advocate monitoring internationally more effectively if it does more itself

For Australia to have an influential role in international agreements, its own monitoring needs to be exemplary. The experience with the migratory wader counts is a good example of how this should be done and what can be achieved. Similarly, Australia is host to the Agreement on the Conservation of Albatrosses and Petrels and is active in advocacy on many fronts to reduce the by-catch of these birds from fishing. The Australian Government regularly funds monitoring of a selection of albatross colonies as well as observers on ships. Through the Antarctic Division it also undertakes research on offal management to reduce albatross deaths and monitors the results.

While waders and albatrosses are relatively easy to count, research is needed to estimate trends in number or habitat condition of many threatened species before monitoring can begin. The southern cassowary (*Casuarius casuarius johnsonii*) is a good example of what has **not** been done. Twenty years after declaration of the Wet Tropics World Heritage Area, the Wet Tropics Management Authority is unable to provide any defensible information on trends in numbers of this, its most high profile threatened species. In its annual reports, upper estimates of the population are said to have declined from 1500 in 2002 to 1200 in 2007 (Wet Tropics Management Authority 2002, 2007), with the population still declining (Wet Tropics Management Authority 2009). While dedicated bureaucrats in the Authority have supported some population research, nowhere in government has this been seen as high priority. In the meantime the only assessments of population are on a small segment of the population using a methodology that assumes foot length of each individual cassowary in an area is unique (Moore 2007), an assertion that remains unproven and improbable. Yet claims of a decline have been used to justify millions of dollars of public expenditure on activities designed to remedy the plight of the southern cassowary; government has no idea whether this investment has had any impact.

Practical solutions

6. Develop metrics to measure skill levels needed by government and those they currently have

It is easy to concentrate monitoring resources on biophysical indicators or on efficient data management and analysis. However, there is also a need to allocate resources to monitor the understanding and use of monitoring results by the public sector. Public sector performance measurement is known to produce positive outcomes (Verbeeten 2008) and it should be included as part of any biodiversity monitoring strategy. For albatrosses and petrels, there is an institutionalised secretariat based in Hobart that ensures the results of any new knowledge are disseminated and employed in negotiations on fishing practice or in the amelioration of other threats. For migratory waders, there has long been a partnership between the Australian Government and the non-government organisations, often leading to support for AWSG initiatives. However, this partnership is not institutionalised, there is no government secretariat dedicated to migratory birds, and the effective translation of monitoring results into effective action has relied on the commitment of individual officers. For threatened species, institutional commitment and appreciation of the need for monitoring varies between jurisdictions. As with migratory species, performance on threatened species appears to be linked to the commitment, interest and influence of individual public servants rather than being a core institutional responsibility. An initial assessment suggests that financial investment in threatened species is much lower where skills turnover is higher. Threatened species also tend to suffer if they stray across tenures (Watson *et al.* 2011) – agencies tend to divide land between protected and non-protected land uses and have trouble categorising (and therefore managing) species that cross these boundaries. Whatever the cause, standardised monitoring of public service performance with respect to monitoring as well as conservation management more generally could be useful, especially if conducted by NGOs.

7. Develop balanced models of government–NGO monitoring partnerships

Good government needs effective measures of progress in all fields of endeavour – economic, social or environmental. However, all forms of monitoring carry with them political risk. Constraints on the budget of the Australian Bureau of Statistics or gradual shifts in the definition of unemployment and inflation could be interpreted as ways of containing that political risk. The same will be true of government-supported biodiversity monitoring. For example, government support for publication of Birds Australia's state of Australia's birds reports has required negotiation on subject and content to accommodate political sensitivities. In contrast to government, non-government organisations can be over-enthusiastic about reporting negative trends (Garnett and Lindenmayer 2011). Thus, while both NGOs and government need each other and government support for biodiversity monitoring attests to the political impact of NGOs, there is an unavoidable tension in reporting. Given this is a universal problem of democratic governance, there is potential to review models of biodiversity reporting elsewhere to understand how tensions can be handled.

The second aspect of the government–NGO relationship is that centralised government-run biodiversity monitoring has the potential to disempower the public who provide both political and practical support for it. There will be a temptation among those with technical skills, particularly for those monitoring aspects of biodiversity for which high levels of technical skill are required for detection and identification, to undertake monitoring using only remote sensing or other techniques that are inaccessible to the public. While this may provide higher levels of accuracy, and will be essential for some measures, there is a need to ensure that the wider public can still contribute to monitoring. Again international models, such as the British Trust for Ornithology monitoring of farmland birds, are worth reviewing.

8. Use some monitoring resources employing people to read and organise published literature

In assessing the status of Australian Important Bird Areas, there was much discussion about using the methods employed in the Atlas of Australian Birds as a national standard for all IBAs. However, it was soon realised that each of the target species that define the eligibility of an area as an IBA require a different method of assessment and a tailored monitoring strategy. The analysis of changes in the IUCN Red List corroborates this observation – almost every threatened species requires a different monitoring technique, and for almost no threatened species is the Atlas of Australian Birds methodology appropriate. The Atlas of Australian Birds data set is irreplaceable for measuring broader trends in avian abundance and diversity but is not designed for threatened species with small ranges. Also, just as there are different monitoring methods, so too there are many ways of reporting the information. Data that allow changes in status to be calculated for the Red List Index are necessarily presented in a wide range of formats and are unlikely to be captured in any efficient web-based data collection platform.

A practical, if non-technical, solution to capture the relevant information would be to use human integrative powers, which still exceed those of computers. Researchers all over the world regularly read and catalogue, either mentally or physically, the latest publications in their field, storing away for future use the information relevant to their research. Similarly, employment of a small number of people to scan intelligently the published and grey literature for data relevant to assessing trends in Australian biodiversity may be an investment that complements more technical and hegemonic biodiversity monitoring regimes. This is essentially what has been done to assess the status of Australian birds every 10 years, each decade's literature being scanned for relevant information. Dedicated literature scanning could ensure not just the Red List Index but a whole range of biodiversity trends are either monitored or validated.

Conclusions

Experience with the use of bird data to monitor Australia's performance against its obligations under a variety of international biodiversity agreements and conventions demonstrates that volunteers have been pivotal. The highly motivated and empowered volunteers with an interest in migratory waders set an example of how monitoring should be undertaken and what can be achieved. Any national biodiversity monitoring strategy needs to engage constructively with this volunteer workforce, not just because they provide data but because they provide the political support for monitoring to take place. Included among this potential volunteer workforce should be those doing research on biodiversity. These researchers do not work voluntarily, in that they are paid to do research, but their data will be effectively 'volunteered' to the monitoring enterprise through their publications and intellectual property. Both community volunteers and biodiversity researchers ought to have a role in designing the strategy so that the vagaries of volunteer and researcher behaviour and motivations are recognised and accommodated, including the multitude of ways in which they provide relevant information. Other countries do this successfully and could provide models.

The potential weak link in any strategy is government, even though the strategy will eventually be delivered by government. Inconsistency in political and financial support as priorities ebb and flow makes it difficult to conduct the long-term monitoring that is essential to track slow environmental trends. Just as important, the degree of skill and understanding among government officers may also be diminishing. The highly accomplished cadre of people who have nurtured so many programs over the last few decades is ageing and few of their replacements appear to have the same background knowledge. This lack of skill may intersect with the tension between government and NGOs on the accessibility to the results of monitoring, or even what is measured and how, that arises from their different political objectives. Finally, there is a danger that government may feel that a national strategy necessarily means creation of a monolithic entity that gathers the same data on every species from every region, failing to appreciate both the variability of biodiversity or of the people and their social institutions that will have responsibility for the monitoring itself. Government support and direction is essential to biodiversity monitoring, and government sanction will be important to incorporate results into policy. But the success of any strategy will be a function of the strength of government's partnership with the passionate public.

Acknowledgements

I am grateful for useful comment on these ideas from Barry Baker, Guy Dutson, Cheryl Gole, Sarah Legge, Danny Rogers, Judit Szabo, David Westcott and John Woinarski. This paper was supported by Australian Research Council Project LP0990395.

Biography

Stephen Garnett is a conservation biologist who has worked extensively with threatened Australian birds in the last two decades. He has spent most of his research life in northern Australia working on natural systems. More recently he has expanded his interests to natural resource-based livelihoods in the region, believing it is only through understanding the motivations of the people inhabiting valuable landscapes that conservation will be achieved.

References

Burger MF and Liner JM (2005). 'Important Bird Areas as a conservation tool: implementation at the State level'. USDA Forest Service General Technical Report PSW-GTR-191.

Dutson G, Garnett ST and Gole C (2009). 'Australia's Important Bird Areas: Key sites for Bird Conservation'. Birds Australia (RAOU) Conservation Statement No. 15.

Garnett ST (1992). 'The Action Plan for Australian Birds'. Australian National Parks and Wildlife Service, Canberra.

Garnett ST and Crowley GM (2000). 'The Action Plan for Australian Birds 2000'. Environment Australia, Canberra.

Garnett ST, Crowley GM and Balmford A (2003). The costs and effectiveness of funding the conservation of Australian threatened birds. *BioScience* **53**, 658–665.

Garnett ST, Crowley GM, Hunter-Xenie HM, Kozanayi W, Sithole B, Palmer C, Southgate R and Zander KK (2009). In the right hands: paying community researchers as an exercise in transformative knowledge transfer. *Biotropica* **41**, 571–577.

Garnett ST and Lindenmayer DB (2011). Conservation science must engender hope to succeed. *Trends in Ecology and Evolution* **26**, 59–60.

Garnett ST, Szabo JK and Dutson G (2011). *The Action Plan for Australian Birds 2010*. CSIRO Publishing, Melbourne.

Geering A, Agnew A and Harding S (2007). *Shorebirds of Australia*. CSIRO Publishing, Melbourne.

Mwangi KMA, Butchart SHM, Munyekenye FB, Bennun LA, Evans MI, Fishpool LDC, Kanyanya E, Madindou I, Machekele J, Matiku P, Mulwa R, Ngari A, Siele J and Stattersfield A J (2010). Tracking trends in key sites for biodiversity: a case study using Important Bird Areas in Kenya. *Bird Conservation International* **20**, 215–230.

Moore LA (2007). Population ecology of the southern cassowary *Casuarius casuarius johnsonii*, Mission Beach north Queensland. *Journal for Ornithology* **148**, 357–366.

Natural Resource Management Ministerial Council (2010). 'Australia's Biodiversity Conservation Strategy 2010–2030'. Australian Government Department of Sustainability,

Environment, Water, Population and Communities, Canberra. <http://www.environment. gov.au/biodiversity/strategy/index.html>

Pullin AS and Knight TM (2005). Assessing conservation management's evidence base: a survey of management-plan compilers in the United Kingdom and Australia. *Conservation Biology* **19,** 1989–1996.

United Nations (2010). 'Millennium Development Goals Report 2010'. United Nations, New York.

United Nations (2011). 'Millennium Development Goals Indicators. Goal 7. Ensure environmental sustainability'. United Nations, New York.

Verbeeten FHM (2008). Performance management practices in public sector organizations: impact on performance. *Accounting, Auditing & Accountability* **21**, 427–454.

Watson JE, Evans MC, Carwardine J, Fuller RA, Joseph LN, Segan DB, Taylor MFJ, Fensham RJ and Possingham HP (2011). The capacity of Australia's protected-area system to represent threatened species. *Conservation Biology* **25**, 324–332.

Wet Tropics Management Authority (2002). 'Appendix 4. State of the Wet Tropics Report 2001–2002'. Wet Tropics Management Authority, Cairns.

Wet Tropics Management Authority (2007). 'State of the Wet Tropics Report 2006–2007'. Wet Tropics Management Authority, Cairns.

Wet Tropics Management Authority (2009). 'State of the Wet Tropics Report 2008–2009'. Wet Tropics Management Authority, Cairns.

5 CHEERFULNESS AND GRUMPINESS IN ECOLOGICAL MONITORING IN AUSTRALIA

Richard J. Hobbs

Lesson #1. Accumulating long-term data sets needs persistence and needn't be expensive.

Lesson #2. Data collected for targeted research questions can be surprisingly useful in contexts other than the original questions.

Lesson #3. Monitoring management activity rather than biodiversity outcomes is easy but largely useless.

Lesson #4. Monitoring is rarely built in as an ongoing, funded part of conservation and restoration projects.

Lesson #5. It's no use collecting data if the information is not going to be available for later use.

Lesson #6. Move beyond vested interests in particular methods and things to be monitored, and focus on what's actually going to help understand the system and aid in its management.

Lesson #7. Build on sites for which current and/or past information is available.

Lesson #8. Build an ability to swing monitoring into action when system-driving events such as fires and floods occur.

Lesson #9. Just do it.

Introduction

It is possible to establish long-term ecological monitoring studies that are focused, produce useful data and have lasting impacts. It seems more difficult, however, to set up effective integrated programs at a national level, at least in Australia. That other places can do it is reason to be cheerful about the prospects for monitoring in Australia. That Australia has, until recently, almost completely failed to establish long-term monitoring systems is, on the other hand, a very good reason to be grumpy. In this chapter I outline, from a personal perspective, some more detailed reasons for cheerfulness and grumpiness, and suggest ways of moving towards more cheerful outcomes in the future. I also pose some questions that require attention as we aim to make this move.

Reasons to be cheerful

1. Accumulating long-term data sets needs persistence and needn't be expensive

Most existing long-term data sets are the result of the efforts of individual researchers setting up and maintaining experiments or sites, rather than institutionalised monitoring schemes. Such long-term data sets frequently produce benefits far outweighing the costs of their collection.

At a conference in 2006 marking the 150th anniversary of the Park Grass Experiment at Rothamsted in England (Silvertown *et al.* 2006), one of the speakers said that there are two rules governing the utility of long-term experiments; first that you can't start them soon enough and second that you can't keep them going long enough. This can be applied to any long-term observations, and yet it seems that institutionalising long-term observations is a hard thing to achieve. There have been discussions in Australia on the development of long-term ecological study and monitoring sites ever since I arrived here in 1984 (and probably for quite some time before that – e.g. Shaughnessy *et al.* 1988), and yet institutionalised arrangements for this are only now beginning to take shape, for instance through the Terrestrial Ecosystems Research Network (TERN – http://www.tern.org.au/).

Although not organised as part of a network of long-term ecological sites in Australia, there are many sites that have been studied over long periods of time, or for which there are historical data. These sites are often the result of the endeavours of one or a few individual scientists and, in relative terms, the cost of their upkeep and continued monitoring is very small. Well-marked and documented plots can be revisited and existing data can be collated and compared with recent information. For instance, in the Western Australian wheat belt, historic bird data were compared with more recent data to reveal long-term trends in bird distributions (Saunders and Ingram 1995), and more detailed repeat site observations yielded very useful information on the long-term trends in ecosystem condition (Saunders *et al.* 2003).

2. Data collected for targeted research questions can be surprisingly useful in contexts other than the original questions

Questions on what to monitor are always easier to answer in the context of specific questions, but some argue that this limits the potential utility of the data later on. However, it is clear that data collected for specific purposes can turn out to be incredibly helpful in broader ways than only in the original context.

Recent assessments of how ecological monitoring can be best carried out have strongly suggested that approaches which focus on specific questions and are based on a clear conceptual model of system behaviour are more likely to yield useful outcomes (Lindenmayer and Likens 2009). Such approaches provide relatively straightforward rationales for decisions on what to measure and remove the necessity of trying to measure everything. Although this may be perceived to limit the likely general utility of the data collected, it is clear from many studies that data collected in one context can actually be very helpful in others. For instance, data from a long-term experiment I started on herbivore exclosure in grassland are now much more valuable for looking at the impacts of climate variability and change on the grassland system (Hobbs *et al.* 2007). Such long-term observations provide the opportunity to capture 'surprise' findings not anticipated at the outset of the study (Lindenmayer *et al.* 2010).

Reasons to be grumpy

3. Monitoring management activity rather than biodiversity outcomes is easy but largely useless

Devising schemes to monitor biodiversity outcomes is hard, and hence often the chosen alternative is the easy option of simply accounting for dollars spent and activities undertaken. This is useful in a simplistic short-term accountability context but completely useless in actually assessing whether the money was well spent and the activity actually worthwhile for biodiversity conservation.

Many government funding schemes and activities run by other organisations have an imperative to spend money wisely and in a timely manner. Often the assessment of whether the money has been spent wisely has to occur very soon after it was spent. What does 'wisely' mean in this context? For instance, if there is a program to fence remnant vegetation against stock grazing, how can the success of the program be judged? From a biological perspective, the success will be measured in terms of recovery of the ecosystem in terms of, for instance, tree regeneration, soil recovery or the presence of particular faunal groups. That involves knowing what the situation was before fencing and being able to observe if this changes subsequently: protocols exist for doing just that and tying the observations to decisions on management (e.g. Brown and Holt 2000).

This seems straightforward enough. However, I have been on several committees associated with funding programs where this has proved challenging, including the Natural Heritage Trust which ran for a decade and spent large amounts of money on biodiversity-related programs. On that committee, monitoring and evaluation was an almost perpetual agenda item that had not been resolved before I left the committee or before the program itself wound up. The debate centred on the tension between trying to find meaningful ways to monitor biological outcomes and having to report on expenditure outcomes in yearly or even half-yearly cycles. Trying to figure out how to determine what the biological outcomes were from hundreds of projects across many different ecosystems conducted by many different types of group proved too difficult, and instead the focus fell back on the relatively straightforward counting of numbers of trees planted, kilometres of fencing installed, numbers of volunteers engaged and so on. As a result, we knew at the end of the program that there had been a truckload of activity. However, we knew virtually nothing about whether this activity had actually made a difference or not.

4. Monitoring is rarely built in as an ongoing, funded part of conservation and restoration projects

Part of the reason Lesson #3 is so prevalent is that biological changes are sometimes slow to take place, and certainly measuring real change requires more than one or two annual funding cycles.

During a workshop run by the WA Department of Conservation and Land Management (now Department of Environment and Conservation) a few years ago, at which experts were gathered to provide input into the development of a forest monitoring system, every expert had a pet process or organism to be measured. Towards the end of the workshop, the facilitator, a senior agency manager, put up an overhead with the likely cost of implementing monitoring to include everything discussed in the workshop – this turned out to be more than the annual operating budget for the whole agency. This, plus the requirement for rapid reporting of outcomes and the lack of ability to carry funds over into longer-term budgets, militates against the inclusion of monitoring processes in projects.

It's also important to consider what sort of monitoring is being discussed. Is it general monitoring of 'condition', or is it targeted monitoring of, for instance, the abundance of a particular species or the spread of a particular invasive organism? Who does the monitoring, and how is it reported? Can we develop efficient means by which front-line managers can carry out regular but efficient observations? Maybe we need to emulate the medical profession, where notification of disease starts with observations by an individual practitioner but are fed into an efficient reporting system that rapidly builds up a picture and provides early warning systems of disease risk. For ecological monitoring, what is the best model: decentralised, on-ground monitoring at one end of the spectrum versus national reporting based on top-down requirements on the other? How can the extremes be pulled together efficiently?

5. It's no use collecting data if the information is not going to be available for later use

There is a wealth of data available from past studies that is useless because the information is not accessible. Publicly funded research should be made available through publication or future access to data when required.

A former PhD student working with me collected data from a nature reserve in Western Australia and we were aware from published work that similar information had been collected in the 1980s, although this was not presented in any detail. We tried numerous avenues for gaining access to the data, but were blocked by the researcher primarily responsible for collecting this information, despite suggesting that the researcher become a collaborator and co-author. Ultimately, we gave up and the student completed her thesis and published the work without reference to the earlier data.

There are numerous reasons past data become unavailable, including simply becoming lost or destroyed, inadequate archiving, inadequate annotation so that the data are largely indecipherable, or being stored in an electronic format which subsequently becomes unreadable. And there is the case outlined above where the holder of the data, for whatever reason, does not want to relinquish the information. This may be because they think they will still publish the data themselves, or they don't want their data misused. For whatever reason, failure to provide access to data is ultimately extremely counterproductive and limits the potential to make the most of historical data in understanding Australia's ecosystems.

Ways to turn grumps to smiles

6. Move beyond vested interests in particular methods and things to be monitored, and focus on what's actually going to help understand the system and aid in its management

There are hundreds of environmental variables and many different species that can be monitored, using many different techniques. Targeting effort on key processes or components streamlines the monitoring process, but requires transparent appraisal of options rather than advocacy of pet methods and organisms.

A few years ago, I attended a workshop that was supposed to be part of the process of setting up a national ecological monitoring program. Rather than focusing on what the monitoring was for and what the questions were, the day quickly disintegrated into a serial display of one vested interest after another: 'Let's use flux towers all over the country'; 'I've got an aeroplane which can fly over areas and measure stuff'; 'We've got access to satellite data which creates pretty colour maps of everything'. Other similar workshops focusing on what groups of organisms should be monitored end up in a similar vein: 'Soil nematodes are a really good indicator of ecosystem degradation'; 'Let's monitor ants because there are lots of them and I like them'; 'Birds are an obvious choice because they're pretty and I can see them without my glasses on'.

Clearly, we need fit-for-purpose monitoring that doesn't try to include everything and isn't infused with vested interests.

7. Build on sites for which current and/or past information is available

From Lesson #1, most existing long-term data sets are the result of the efforts of individual researchers setting up and maintaining experiments or sites, rather than institutionalised monitoring schemes. These long-term studies can form the kernel of a national network of long-term sites.

Recognition of existing sites could yield much needed stability and continuity to studies that might otherwise cease when the principal researcher runs out of funds, retires or passes on. A relatively small amount of funding, efficiently dispensed, could achieve this. However, in attempting to draw such sites and studies into some sort of national framework, the trick will be to do so in such a way that the individual endeavours are not swamped and stifled by a new bureaucracy which imposes strict reporting, unnecessarily uniform standards and so on. Is it possible to mesh together diverse studies in such a way that they add to more than the sum of the individual parts? Further, is it possible to do this without creating a layer of administration that sucks most of the funding away from the actual monitoring processes?

8. Build an ability to swing monitoring into action when system-driving events such as fires and floods occur

Although much change happens slowly, there are occasions where rapid change occurs and system responses may be both massive and critically important in the long-term dynamics of the system. Whatever monitoring schemes we devise have to be nimble enough to be able to respond to these occasions.

9. Just do it

Pushing the topic of long-term monitoring around the table for decades keeps people busy but doesn't actually achieve much. Surely we can be smart enough to overcome the foibles of the past and develop a useful way forward.

Biography

Richard Hobbs is Professor of Restoration Ecology in the School of Plant Biology at the University of Western Australia, where he holds an ARC Australian Laureate Fellowship, and leads the Ecosystem Restoration and Intervention Ecology Research Group. Originally from Scotland, he spent three years in California and has been in Western Australia since 1984, working with CSIRO and at Murdoch University before joining UWA in 2009. His particular interests are in vegetation dynamics and management, invasive species, ecosystem restoration, conservation biology and landscape ecology. He has several long-term ecological studies under way including a 28-year study of California grassland dynamics.

References

Brown L and Holt C (2000). 'Eucalypt woodlands: a guide to management'. Miscellaneous Publication 17/00. Agriculture Western Australia, WA.

Hobbs RJ, Yates S and Mooney HA (2007). Long-term data reveal complex dynamics in grassland in relation to climate and disturbance. *Ecological Monographs* **77**, 545–568.

Lindenmayer DB and Likens GE (2009). Adaptive monitoring: a new paradigm for long-term research and monitoring. *Trends in Ecology and Evolution* **24**, 482–486.

Lindenmayer DB, Likens GE, Krebs CJ and Hobbs RJ (2010). Improved probability of detection of ecological 'surprises'. *Proceedings of the National Academy of Sciences of the USA* **107**, 21957–21962.

Saunders DA and Ingram J (1995). *Birds of Southwestern Australia. An Atlas of Changes in the Distribution and Abundance of the Wheatbelt Fauna.* Surrey Beatty and Sons, Chipping Norton.

Saunders DA, Smith GT, Ingram JA and Forrester RI (2003). Changes in a remnant of salmon gum *Eucalyptus salmonophloia* and York gum *E. loxophleba* woodland, 1978 to 1997. Implications for woodland conservation in the wheat–sheep regions of Australia. *Biological Conservation* **110**, 245–256.

Shaughnessy PD, Leigh JH and Walker BH (1988). Preliminary proposals for a set of long-term ecological study and monitoring sites in Australia. Unpublished Report. CSIRO Division of Wildlife and Ecology, Canberra.

Silvertown J, Poulton P, Johnston E, Edwards G, Heard M and Bliss PM (2006). The Park Grass Experiment 1856–2006: its contribution to ecology. *Journal of Ecology* **94**, 801–814.

6 THE CONSERVATION RETURN ON INVESTMENT FROM ECOLOGICAL MONITORING

Hugh P. Possingham, Brendan A. Wintle, Richard A. Fuller and Liana N. Joseph

Lesson #1. Most ecological monitoring starts without clearly stating the purpose of that monitoring, so our chapter defines five clear reasons to monitor.

Lesson #2. Money spent on monitoring is money that is not spent on management – there is a trade-off – and in some cases it may not be worth monitoring conservation management actions at all.

Lesson #3. Once the purpose of monitoring has been determined, the expected benefits and costs of that monitoring must be quantified so that a conservation return on investment can be calculated.

Lesson #4. There are at least five distinct, potentially quantifiable, benefits of monitoring.

Lesson #5. Monitoring optimised for one benefit may not be optimal for other benefits, so we need to focus on the primary objective or adopt a broad multi-criteria approach to determine the relative merits of different monitoring proposals.

Lesson #6. We need a new field of research that predicts, and test predictions about the expected benefits of ecological monitoring.

Introduction

Ecological monitoring is firmly established as an essential tool for the effective management of biodiversity (Strayer *et al.* 1986). Many papers and books describe the reasons why ecological monitoring is important and outline the principles for good monitoring (e.g. Lindenmayer and Likens 2010). In this chapter, we will look at the more controversial question of whether all biodiversity management interventions require monitoring, and if so, what fraction of our effort should be put into that monitoring. We pay particular attention to long-term 'surveillance' monitoring, as this has been criticised in the literature as generally not delivering value for money (Nichols and Williams 2006).

While there are compelling reasons to undertake ecological monitoring, those reasons are not always clearly articulated. Monitoring is likely to play an increasingly important role as we try to determine how large-scale anthropogenic changes, in the context of a rapidly changing climate, alter the way in which we should manage ecosystems. Furthermore, effective monitoring is a critical link in the cycle of adaptive management that aims to iteratively improve conservation actions over time (Holling 1978; Walters 1986) and is important for influencing government policy and investment in environment programs.

Lessons

1. Most ecological monitoring starts without clearly stating the purpose of that monitoring, so our chapter defines five clear reasons to monitor

What is monitoring for? We believe there are five distinct conservation, scientific and social benefits of long-term monitoring (Yoccoz *et al.* 2001; Nichols and Williams 2006; Salzer and Salafsky 2006; McDonald-Madden *et al.* 2010a): (1) monitoring the state of a system so that we can make an appropriate state-dependent management response – such as harvesting a certain number of animals; (2) active adaptive management – the process where we sacrifice short-term gains in management to learn about a system so that we can make better decisions in the future; (3) finding out about how biodiversity is faring so we can alert the public and politicians to unwanted change – awareness raising; (4) engaging the public in ecological issues thereby leveraging effort and support, such as reporting on declines in Australian woodland birds; and (5) discovering useful things we did not expect – serendipity. We briefly discuss these five benefits of monitoring with a particular focus on quantifying their benefits.

1. Monitoring the state of a system in order to choose the best management actions
Much of our best long-term ecological data comes through monitoring of harvestable animals (Grigg *et al.* 1999; Fisheries Queensland 2010). Fisheries stock assessment relies on data collected by fishers to determine next season's effort and/or quota (Clark and Mangel 2000; Dichmont *et al.* 2010). In Australia, the annual kangaroo harvesting is driven by aerial surveys, and the annual waterbird counts were initiated to determine annual duck harvesting regulations (Kingsford and Thomas 2004; Pople *et al.* 2007).

In all of these situations, exactly how much we choose to harvest is determined directly by the monitoring data (Johnson and Williams 1999). These monitoring programs were driven by the utilitarian needs of prudent management but have provided many other benefits, including a deeper understanding of the population ecology of the species concerned (e.g. Jonzén *et al.* 2002, 2005). Similarly, some of our better long-term monitoring data sets on vegetation state and fauna population dynamics were driven by the needs of fire managers (e.g. Keith *et al.* 2007; Penman *et al.* 2007).

State-dependent management demands long-term, continuous ecological monitoring so that changes in system state can be reliably identified and acted upon. While state-dependent management is our most successful driver of long-term ecological monitoring, few people have questioned exactly how precisely we need to track the state of a system to make good management choices. To establish how much to spend on monitoring systems where management is state-dependent, we must quantify the net benefit of accurately knowing the system state. Specifically, we need to know what cost we would pay if a poor estimate of population size or ecological state caused us to make a poor decision (Hauser *et al.* 2006) or how the marginal benefits of monitoring (in terms of good management decisions) change with increased investment in monitoring. For example, capturing individuals of a rare species for captive breeding when they are not in decline will increase the extinction probability of the population and cost money (Tenhumberg *et al.* 2004). On the other hand, if we intervene too late, it will be more expensive, or impossible, to secure the tiny population we have. As investment in monitoring increases, false alarm and overconfidence errors should begin to decrease. However, there is a point at which further investment is not worthwhile, given the expected costs of these errors. In summary, each type of mistake has costs and benefits that need to be traded off against the investment in monitoring (Field *et al.* 2004; Hauser *et al.* 2006).

2.   Active adaptive management: monitoring to learn about a system
Active adaptive management is learning by doing (Shea *et al.* 1998). While state-dependent management uses monitoring to learn the state of the system, active adaptive management thinking recognises that we are often also uncertain about how the system works. This can be characterised as uncertainty about system parameters or processes such as: estimates of a species' birth and death rates, the influence of environmental variables on carrying capacity, or the degree to which management actions influence a particular state variable. Learning is essential to the management process, both for understanding the influence of particular management actions or strategies, and for learning about natural processes or parameters of models that may be used to choose between management options (Drechsler and Burgman 2004; Gerber *et al.* 2005; Nichols and Williams 2006).

The key to the trade-off between learning and managing is knowing when the costs of long-term gains from more learning are going to be compensated for by better conservation management outcomes (McCarthy and Possingham 2007). The costs and benefits of the investment in learning can be traded off with each other to find the optimal level of monitoring. This is precisely the approach used by Gerber *et al.* (2005), who discovered that the optimal monitoring period is rarely longer than five years when monitoring the recovery of an over-fished stock to set the fraction of that stock to protect. So when would learning-based monitoring need to be long term, say more than a decade?

If the fundamental parameters of an ecological system are constant, then monitoring for learning will rarely need to be more than a decade. This said, some fundamental parameters inherently require much more than a decade to learn, for example the frequency of very rare events. Furthermore, the more we learn about a system, invariably the more we discover we do not know. From the perspective of gaining such new knowledge, monitoring could go on indefinitely, although one might expect the rate of gaining new knowledge to decrease. For most practical problems, sufficient knowledge to address those problems will usually be acquired in less than a decade depending on the cost of the monitoring relative to the expected practical benefits with regard to decision-making. However, the generalisations about diminishing returns of monitoring with time may not apply in systems undergoing major and continuous change. The role of active adaptive management as a tool for updating knowledge and mediating responses to rapid environmental change is likely to be significant.

3.   Awareness: monitoring to raise awareness of an issue for the public and policy makers
Long-term monitoring has a particularly important role in informing the public, and decision makers, about long-term changes in our ecosystems and species. This is especially important because of the phenomenon known as 'shifting baselines' whereby it is hard to perceive important long-term change because observers of a system alter their internal point of reference to reflect recent conditions (Jackson *et al.* 2001). This is particularly important in a country such as Australia where rare events and climate cycles with a long period seem to play a large role in shaping ecological systems, in comparison with Europe and North America where the dominant environmental cycle for most systems is annual.

There are some outstanding examples of long-term monitoring programs that have fundamentally changed the way society thinks about ecological systems and the state of their environment. Long-term monitoring of changes in climate, the ozone layer and the acidity of rain, have led to huge (although not always adequate and timely) changes in policy and management (Likens *et al.* 1996). These issues have demanded long-term monitoring and dedicated persistent scientists. Without compelling data and analysis, society would not have made the necessary and frequently expensive responses. The benefits of these long-term monitoring programs are unquestionable, and far exceed their costs. However, for some ecological problems, the evidence for a problem may already be large, the benefits of monitoring may be much smaller and the chance of gathering sufficient long-term data in time to make a useful decision may be slim. A better choice in such situations may be to use resources to take action without bothering to collect data to establish that there is a problem (Field *et al.* 2004).

4.   Engagement: monitoring to engage the public and leverage effort
Garnering public and political support may result from two mechanisms: (1) publicising the outcomes of a monitoring program (Whitelaw *et al.* 2003) as described above, or (2) through engagement in the process of monitoring (Lindenmayer *et al.* 1991; Vaughan *et al.* 2003). Both of these mechanisms may influence management because policy makers often respond to publicity about detrimental environmental trends and/or pressure from the public (Cuthill 1995; McNeil *et al.* 2006).

Monitoring schemes that have the objective of increasing awareness through participation include government-sponsored, frequently volunteer-driven, monitoring programs mandated by some level of local, regional or national government (e.g. community-based monitoring efforts; Vaughan *et al.* 2003; Whitelaw *et al.* 2003; Trewhella *et al.* 2005). They are useful for educating the community about local conservation issues and forming a vehicle for government agencies to provide the public with information about the status of the natural environment. These programs are often motivated by the need to involve stakeholders and citizens in planning and management processes (Cuthill 1995) to ensure social, cultural and economic perspectives are reflected in evidence-based decisions about conservation priorities and actions, and to encourage conflict resolution and advance social learning (Fernandez-Gimenez *et al.* 2008). However, the cost-efficiency of these monitoring programs should be considered carefully; community-based monitoring may be an expensive method of raising awareness for only a small sector of society that is already well engaged. Managers should consider how limited resources should be most efficiently used to educate and influence more people (e.g. field days, school outings and publicity campaigns).

Monitoring programs with the aim of increasing awareness may or may not be used to meet one or more of the other monitoring objectives. In fact, awareness-raising objectives may conflict with other monitoring objectives because more complex or tedious monitoring programs may result in more useful data to detect trends, but may result in fewer people participating in the data collection. The designers of education-focused monitoring programs need to be clear about the quality and potential limits of the data collected to maintain participant confidence and avoid triggering inappropriate or unnecessary management actions based on data of poor quality or from poorly designed collection programs (Hockley *et al.* 2005). The designers of such programs should be clear about the tangible benefits that are expected to arise from the investment in community education, and question whether these outweigh the benefits of investing in a more scientifically rigorous monitoring scheme or just taking action in the absence of community support.

Community monitoring is often free, and indeed may even leverage substantial additional resources that can be used for management. Remarkably, many people just want to monitor, and some will even pay to do so. Crucially though, the availability of a cheap source of labour must not be seen as a reason to monitor. People seem especially to enjoy untargeted monitoring to just see what is happening in the environment and such monitoring is best perceived as a bonus. However, if the existence of a community-monitoring program with expectations about the probability of providing good scientific results reduces the motivation or impetus to invest in a targeted monitoring program, then they may be counterproductive.

5.   Serendipity

Some authors are critical of untargeted monitoring (Anderson 2001; Nichols and Williams 2006; Lindenmayer and Likens 2010) – monitoring that is possibly intended to uncover the 'unknown unknowns'. We agree that it is hard to determine how much effort we should allocate to this sort of monitoring, especially since the chance and value of discovering something unexpected is hard to quantify (Wintle *et al.* 2010). However, it is true that serendipitous discoveries, such as the impact of pesticides on North American birds (Carson 1962) have led to far-reaching conservation benefits by driving policy reform and increasing public interest. We perceive two kinds of serendipity in long-term ecological monitoring. First, there are the serendipitous gains simply by having people in the environment and looking, which is precisely why Lindenmayer and Likens (2010) argue that even the old chief investigators should get out of their armchairs once in a while and actually count something. Interesting things happen in nature and you have to be there to see them. Second, there are the unexpected discoveries that emerge simply from the age of a data set – events we couldn't have predicted but that accumulate as a data set matures. In many instances, the serendipitous discoveries arising from monitoring are unrelated, or only tenuously related to the aim of the monitoring program (Wintle *et al.* 2010). For example, the identification of the acid rain phenomenon (Likens *et al.* 1972; Likens and Bormann 1974) arose from data collected as part of an unrelated (long-term) study of catchment-level nutrient budgets in the Hubbard Brook watershed (Likens *et al.* 1970). It is conceivable that a large-scale post-hoc evaluation of existing long-term monitoring programs could be used to evaluate the probabilistic benefits of serendipitous monitoring (Wintle *et al.* 2010). For example, one could empirically evaluate what has been discovered accidentally and at what cost, as well as determine the action that was catalysed by those discoveries.

2. Money spent on monitoring is money that is not spent on management – there is a trade-off – and in some cases it may not be worth monitoring conservation management actions at all

Conservation budgets are finite, and money spent monitoring is money not spent on other biodiversity conservation action. So, despite increasing interest and investment in ecological monitoring, the problem is that these resources have to come from somewhere. Information gain is not necessarily conservation gain. Monitoring is widely regarded as such a rational and defensible activity in the pursuit of improved conservation outcomes that it usually has a default place within any conservation project. Rarely, however, do we critically assess the relative value of gaining information. We must be prepared to forgo monitoring in some cases by explicitly asking the question: Is spending money on monitoring justified relative to funding other actions, including strategic research? In other words, we must trade off investment in monitoring with investment in management and be receptive to the idea that in many cases, despite the potential benefits of monitoring, the best choice in some circumstances could be not to invest in monitoring at all (McDonald-Madden *et al.* 2010a).

3. Once the purpose of monitoring has been determined, the expected benefits and costs of that monitoring must be quantified so that a conservation return on investment can be calculated

Choices about if and what to monitor can be cast as a resource allocation problem – monitor or manage; monitor what within a project; what kind of monitoring is best; do some projects need more monitoring than others; can we make some conservation interventions without monitoring their impact at all? Globally, only a small proportion of the species threatened with extinction are monitored (Baillie *et al.* 2004) and, in many cases, the monitoring is insufficient to detect real changes in parameters and responses of populations and/or ecosystems to management intervention (Taylor and Gerrodette 1993; Field *et al.* 2004). Consequently, we need to choose among the many aspects of an ecological community or population that we could monitor, and the techniques we use to monitor them. Essentially, we must decide how to allocate resources among alternative possible long-term ecological monitoring programs to maximise conservation outcomes. To answer all of these questions we need to be able to calculate the conservation return on investment from monitoring.

To determine a conservation return on investment from monitoring we need to quantify two things: costs and benefits. First, we need to know the current and expected cost of a monitoring program. While this is rarely reported in the conservation literature, it is a task that is not very complex, and we will assume that any agency with good accountants can calculate the expected total cost of any monitoring program. Second, we need to determine the expected benefits of an ecological monitoring program. Calculating benefits is much more difficult because there is no single natural currency. If dollars is the natural currency for calculating costs, what is the natural currency for measuring biodiversity benefits? There are a few good candidates that are consistent with the fundamental objectives of most biodiversity conservation programs, for example:

- the decrease in extinction probability following management improved because of monitoring;
- an increase in population growth rate of a keystone species because we make state-dependent interventions in a more timely fashion; and
- the reduction in the costs of managing other biodiversity conservation projects because a given monitoring program delivered knowledge useful for those other projects.

Unfortunately, none of these benefits has the same currency. The first benefit is measured as a probability, the second benefit is measured as a rate of population growth, and the third is measured in dollars. Even in the case of the third measure, a biodiversity benefit measure must exist to determine the usefulness of knowledge to other conservation programs. Adding to the confusion about currencies of biodiversity benefit is the fact that most of these benefits are quite difficult to measure retrospectively and even more difficult to predict *a priori*. Despite these apparently overwhelming obstacles to quantifying benefit, we will begin with the essential first step of classifying the different kinds of benefit arising from ecological monitoring. Ironically, identifying the expected benefits of a monitoring program is a step that most people embarking on monitoring program design do not take.

4. There are at least five distinct, potentially quantifiable, benefits of monitoring

We have now explored the factors that we need to think about in terms of the costs and benefits of monitoring – but how does this help us make decisions about how much and what to monitor? For the simpler cases of optimal harvesting (Clark and Mangel 2000; Hauser *et al.*

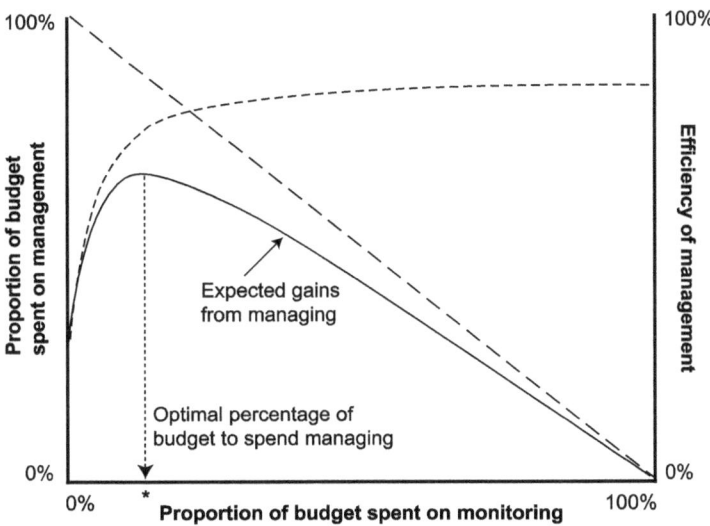

Figure 6.1. The optimal allocation of resources to monitoring and management given a fixed total budget for both. The heavy dashed line represents the conservation outcome achieved with 100% efficiency (i.e. no error due to lack of understanding of state or system), given that a proportion of the budget is spent on monitoring. This line declines with a slope of –1 because more money that is spent on monitoring, the less is available for management. The light dashed line represents the improvement in management efficiency gained from monitoring given the budget spent on monitoring. We assume that management efficiency starts low and increases rapidly as we invest, but returns rapidly diminish. The solid line represents the net outcome from management given a proportion of funding is used for monitoring. As long as the slope of the management efficiency curve is greater than one, then there will be a non-zero optimal proportion of the budget, *, that should be spent on monitoring.

2006) and optimal learning (Gerber *et al.* 2005; Grantham *et al.* 2009; McDonald-Madden *et al.* 2010b) such analyses are relatively well advanced, and solutions to the optimal monitoring problem are available. In each case, we can calculate the increase in management efficiency from investment in monitoring and trade that off against the loss in management effort (see Figure 6.1). In theory, there will be an optimal amount to monitor that depends on factors such as our current state of knowledge, the costs of different kinds of ignorance (ignorance about the system state and ignorance about how the system works) and the cost of the monitoring itself. While this is rarely done, and few good examples exist, the bigger challenge is to determine how much to monitor when the benefits of monitoring are dominated by the more nebulous gains from informing opinion, engaging the public and serendipitous discoveries. In each case, the three kinds of benefits will ebb and flow, often unpredictably. However, while some authors have argued that the last three reasons to monitor are vague and hard to quantify, that does not mean they are of no value. We believe that they do have value and there is an urgent need to quantify the benefits of monitoring that precipitates political action, engages the public and leads to unexpected discoveries. Below we speculate how the benefits of these three kinds of monitoring might change through time.

Initially, the ability of a monitoring program to identify patterns of concern and inform policy or engage the public will tend to be low. Two or three years of data is not enough to compel even the most naïve member of the public, and is usually statistically inadequate to detect change in response to managed or unmanaged processes. Hence, raising public concern over an issue will take time. Conversely, the public's enthusiasm for all things new may lead to strong early engagement with volunteers. Over five to 15 years much of the most dramatic and important information about how biodiversity is changing will have been uncovered. These will be the discoveries that we expect to make, usually the main drivers for the study in the first place. For example, are Australian waterbirds in decline from over-harvesting of water, how is the extinction debt playing out for woodland fauna? After that, the rate of acquiring new information about long-term change will decline, although if some of the important environmental events are rare and/or have long cycles, valuable information could continue to emerge for decades. Once a monitoring program has existed for a long period of time, probably measured in decades, public engagement may rise again as people see the power of a long-term data set.

But what about serendipity – the ability of long-term monitoring data to uncover previously unforeseen phenomena (Wintle *et al.* 2010)? We argue that there are two kinds of serendipitous discovery, those that occur simply because we are observing a system or a place that has not previously been monitored. This will be greatest early in the program and depend on how well studied the system was beforehand. Then there are the serendipitous discoveries resulting from unexpected events, and many of those will only emerge after a long period of time, or when the data are amalgamated with other information in unforseen ways (Lindenmayer *et al.* 2010). There is a third, indirect role for long-term monitoring data in serendipitous discovery. This is the role of corroborating a new idea that has arisen by means other than the monitoring data. There have been at least a couple of notable examples in which monitoring data has played a key role in corroborating a new idea or hypothesis. Wintle *et al.* (2010) describe the case in which a long-term, 'unfocused' monitoring program (roadside spotlighting of mammals in Tasmania) was used in a post-hoc manner to support the hypothesis that Tasmanian Devil Facial Tumour disease (DFTD) was impacting on the abundance of Tasmanian Devils (*Sarcophilus harrisii*; Hawkins *et al.* 2006). In this case, the monitoring data, collected for a different purpose (regulation of wallaby culling), were not involved in the discovery of DFTD or even in generating concern about population level consequences, but they were central to demonstrating the catastrophic effect of the disease on the local population and motivating urgent political action.

We postulate that the 'hard to quantify' benefit curves of long-term ecological monitoring will generally have a shape something like those in Figure 6.2, though the absolute magnitude of each kind of benefit will vary from case to case. One thing we can be sure of – these benefits are real and only careful retrospective analyses of long-term ecological monitoring programs will reveal credible patterns in, and understanding of, how these benefits change through time.

5. Monitoring optimised for one benefit may not be optimal for other benefits, so we need to focus on the primary objective or adopt a broad multi-criteria approach to determine the relative merits of different monitoring proposals

Given these five disparate benefits of monitoring, it is clear that a monitoring strategy optimised for one benefit will not necessarily deliver the others efficiently. For example, an ecological monitoring program designed to educate the public and politicians might focus on systems or species that people particularly care about (e.g. charismatic mammals or birds), and the subject of awareness-raising monitoring may be chosen simply to encourage participation. Monitoring the charismatic species that people care most about may be the best way to maximise community interest in conservation, but may not be the best way to learn about the functioning of an ecosystem or to measure the state of a resource (habitat or prey abundance) critical to the survival of that species. The monitoring strategy that will provide the necessary information for making good decisions about habitat management may focus on other species or habitat attributes that are less charismatic, but a more reliable and cost-effective means of learning and tracking system state (Lindenmayer *et al.* 2000). Therefore, it is vitally important

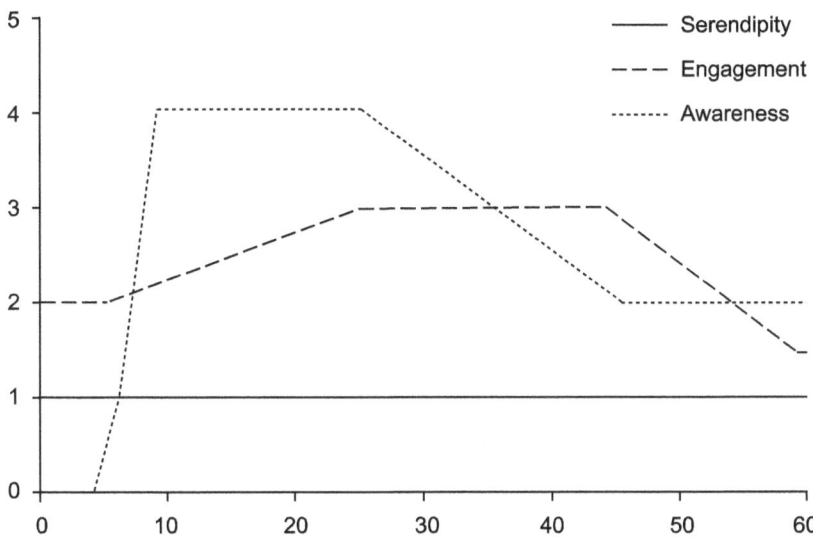

Figure 6.2. How three of the benefits of ecological monitoring might change through time as the ecological monitoring program progresses. Serendipity is a combination of chance observations, which may be high early as we explore a new system in a new way, and uncovering long-term and unexpected changes. Awareness won't increase until we have enough data to make credible claims and engagement is likely to lag awareness, although may start relatively high generated by a new program. We know very little about the real shape of these benefit curves.

that the design of a monitoring program be predicated on a very clear understanding of exactly what the primary objective of the monitoring program is, even if incidental benefits are antici-pated. In some cases we may need to adopt a broad multi-criteria approach to assessing the relative benefits of monitoring programs.

6. We need a new field of research that predicts, and test predictions about the expected benefits of ecological monitoring

While exact quantification of the likely costs and benefits of ecological monitoring is difficult, there is no other rational way to allocate effort within and between ecological monitoring programs. In the absence of any quantification of costs and benefits, allocation of funding to monitoring will continue based on unsubstantiated gut feeling. One of the greatest challenges in ecology and conservation biology is in specifying and quantifying the benefits of long-term ecological monitoring programs in a way that can be empirically tested.

In this chapter, we have classified the benefits of ecological monitoring and reviewed examples of attempts to quantify some of these benefits, particularly for state-based decision-making and learning. There is, however, an urgent need for robust techniques for quantifying some of the less well understood benefits to biodiversity of monitoring, for example engaging the public to support and conduct conservation, informing policy makers, or serendipitous discoveries, such as uncovering novel or unpredictable threats or remedies. The last of these benefits of monitoring will be the most difficult to quantify, although we do not believe it impossible.

To ensure that we invest in the best ecological monitoring programs and to ensure ongoing investment in monitoring, it is imperative that we compare the costs and benefits of different options. To do this, we need to calculate the costs and predict the expected benefits including the ability to make state-dependent decisions, learn about a system, inform the public or policy makers, engage the public or be sure that novel and unpredictable events are not overlooked.

Although research on quantifying these benefits is nascent, we are beginning to see approaches and case studies illustrating how to do this. As the fields of optimal monitoring and adaptive management continue to progress, it will become possible to operationalise the framework that we have presented here to identify which ecological monitoring programs are mostly likely to deliver the best biodiversity outcomes in a cost-constrained world.

Acknowledgements

This research was largely funded by grants from the Australian Research Council and a Commonwealth Environmental Research Facility grant from the Department of the Environment, Water, Heritage and the Arts. We are grateful to David Lindenmayer and Max Bourke for inviting us to contribute to this volume and inspiring more rigorous thinking about ecological monitoring.

Biographies

Hugh Possingham is the Director of the Australian Research Council Centre of Excellence for Environmental Decisions. Since 2000 he has been a Professor of Mathematics and Professor of Ecology at The University of Queensland. He is a member of the Council of The Australian Academy of Science, has won two Eureka prizes and the inaugural Fenner medal. He has published over 350 peer-reviewed papers. Hugh also wrote the 'Brigalow Declaration' which was used by Premier Beattie to stop land clearing in Queensland. His main area of research is using decision science methods to solve environmental problems.

Brendan Wintle is a Senior Lecturer at the University of Melbourne's School of Botany where he holds an ARC Future Fellowship, is the Deputy Director of the National Environmental Research Program Decisions Hub, and theme Leader in the ARC Centre of Excellence for Environmental Decisions. He worked in forest policy for the Queensland Government before completing his PhD. Brendan specialises in ecological modelling and uncertainty in environmental decision-making and writes on technical and policy issues concerned with conservation.

Richard Fuller is a lecturer in biodiversity and conservation at The University of Queensland. His main research interests are in conservation planning, urban ecology, and migratory species conservation. Recent projects include working with the Australian Government to understand and reverse the declines in migratory shorebirds, and planning how to adapt protected area systems to the future threats of landscape clearance and climate change.

Liana Joseph is a David H. Smith Conservation Research Fellow at the Wildlife Conservation Society in New York. Her main research interest is the conservation of endangered species. Past projects have included working with the New Zealand and Australian governments to design management strategies for threatened species. Her most recent research focuses on the over-exploitation of wildlife and examining the effectiveness of conservation strategies for protecting species threatened by trade.

References

Anderson DR (2001). The need to get the basics right in wildlife field studies. *Wildlife Society Bulletin* **29**, 1294–1297.

Baillie JEM, Hilton-Taylor C and Stuart SN (2004). '2004 IUCN Red List of threatened species: A global species assessment'. IUCN, Gland, Switzerland and Cambridge, UK.

Carson R (1962). *Silent Spring*. Houghton Mifflin, Boston, Massachusetts, USA.

Clark CW and Mangel M (2000). Conservation Biology. In *Dynamic State Variable Models in Ecology: Methods and Applications*. (Eds WC Clark and M Mangel) pp. 173–191. Oxford University Press, Oxford, UK.

Cuthill M (1995). An interpretive approach to developing volunteer-based coastal monitoring programmes. *Local Environment* **5**, 127–137.

Dichmont CM, Pascoe S, Kompas T, Punt AE and Deng R (2010). On implementing maximum economic yield in commercial fisheries. *Proceedings of the National Academy of Sciences of the USA* **107**, 16–21.

Drechsler M and Burgman MA (2004). Combining population viability analysis with decision analysis. *Biodiversity and Conservation* **13**, 115–139.

Fernandez-Gimenez ME, Ballard HL and Sturtevant VE (2008). Adaptive management and social learning in collaborative and community-based monitoring: A study of five community-based forestry organizations in the western USA. *Ecology and Society* **13**, 4.

Field SA, Tyre AJ, Jonzén N, Rhodes JR and Possingham HP (2004). Minimizing the cost of environmental management decisions by optimizing statistical thresholds. *Ecology Letters* **7**, 669–675.

Fisheries Queensland (2010). 'Annual status report 2010: Rocky reef fish fishery'. Department of Employment, Economic Development and Innovation, Brisbane, Australia.

Gerber LR, Beger M, McCarthy MA and Possingham HP (2005). A theory for optimal monitoring of marine reserves. *Ecology Letters* **8**, 829–837.

Grantham HS, Wilson KA, Moilanen A, Rebelo T and Possingham HP (2009). Delaying conservation actions for improved knowledge: How long should we wait? *Ecology Letters* **12**, 293–301.

Grigg GC, Beard LA, Alexander P, Pople AR and Cairns SC (1999). Aerial survey of kangaroos in South Australia 1978–1998: A brief report focusing on methodology. *Australian Zoology* **31**, 292–300.

Hawkins CE, Baars C, Hesterman H, Hocking GJ, Jones ME, Lazenby B, Mann D, Mooney N, Pemberton D, Pyecroft S, Restani M and Wiersma J (2006). Emerging disease and population decline of an island endemic, the Tasmanian devil *Sarcophilus harrisii*. *Biological Conservation* **131**, 307–324.

Hauser CE, Pople AR and Possingham HP (2006). Should managed populations be monitored every year? *Ecological Applications* **16**, 807–819.

Hockley NJ, Jones JPG, Andriahajaina FB, Manica A, Ranambitsoa EH and Randriamboahary JA (2005). When should communities and conservationists monitor exploited resources? *Biodiversity and Conservation* **14**, 2795–2806.

Holling CS (1978). *Adaptive Environmental Assessment and Management*. Blackburn Press, Caldwell, NJ.

Jackson JBC, Kirby MX, Berger WH, Bjorndal KA, Botsford LW, Bourque BJ, Bradbury RH, Cooke R, Erlandson J, Estes JA, Hughes TP, Kidwell S, Lange CB, Lenihan HS, Pandolfi JM, Peterson CH, Steneck RS, Tegner MJ and Warner RR (2001). Historical overfishing and the recent collapse of coastal ecosystems. *Science* **293**, 629–637.

Johnson F and Williams K (1999). Protocol and practice in the adaptive management of waterfowl harvests. *Conservation Ecology* **3**, 8.

Jonzén N, Lundberg P, Ranta E and Kaitala V (2002). The irreducible uncertainty of the demography-environment interaction in ecology. *Proceedings of the Royal Society of London B* **269**, 221–225.

Jonzén N, Pople AR, Grigg GC and Possingham HP (2005). Of sheep and rain: Large-scale population dynamics of the red kangaroo. *Journal of Animal Ecology* **74**, 22–30.

Keith DA, Holman L, Rodoreda S, Lemmon J and Bedward M (2007). Plant functional types can predict decade-scale changes in fire-prone vegetation. *Journal of Ecology* **95**, 1324–1337.

Kingsford RT and Thomas RF (2004). Destruction of wetlands and waterbird populations by dams and irrigation on the Murrumbidgee River in arid Australia. *Environmental Management* **34**, 383–396.

Likens GE and Bormann FH (1974). Acid rain: A serious regional environmental problem. *Science* **184**, 1176–1179.

Likens GE, Bormann FH and Johnson NM (1972). Acid rain. *Environment* **14**, 3–40.

Likens GE, Bormann FH, Johnson NM, Fisher DW and Pierce RS (1970). Effects of forest cutting and herbicide treatment on nutrient budgets in the Hubbard Brook Watershed-Ecosystem. *Ecological Monographs* **40**, 23–47.

Likens GE, Driscoll CT and Buso DC (1996). Long-term effects of acid rain: Response and recovery of a forest ecosystem. *Science* **272**, 244–246.

Lindenmayer DB, Tanton M, Linga T and Craig S (1991). Public participation in stagwatching surveys of a rare mammal: Applications for environmental and public education. *Australian Journal of Environmental Education* **7**, 63–70.

Lindenmayer DB, Margules CR and Botkin DB (2000). Indicators of biodiversity for ecologically sustainable forest management. *Conservation Biology* **14**, 941–950.

Lindenmayer DB and Likens GE (2010). *Effective Ecological Monitoring.* CSIRO Publishing, Melbourne.

Lindenmayer DB, Likens GE, Krebs CJ and Hobbs RJ (2010). Improved probability of detection of ecological 'surprises'. *Proceedings of the National Academy of Sciences of the USA* **107**, 21957–21962.

McCarthy MA and Possingham HP (2007). Active adaptive management for conservation. *Conservation Biology* **21**, 956–963.

McDonald-Madden E, Baxter PWJ, Fuller RA, Martin TG, Game ET, Montambault J and Possingham HP (2010a). Monitoring does not always count. *Trends in Ecology and Evolution* **25**, 547–550.

McDonald-Madden E, Probert WJM, Hauser CE, Runge MC, Possingham HP, Jones ME, Moore JL, Rout TM, Vesk PA and Wintle BA (2010b). Active adaptive management of threatened species in the face of uncertainty. *Ecological Applications* **20**, 1476–1489.

McNeil TC, Rousseau FR and Hildebrand LP (2006). Community-based environmental management in Atlantic Canada: The impacts and spheres of influence of the Atlantic coastal action program. *Environmental Monitoring and Assessment* **113**, 367–383.

Nichols JD and Williams BK (2006). Monitoring for conservation. *Trends in Ecology and Evolution* **21**, 668–673.

Penman TD, Kavanagh RP, Binns DL and Melick DR (2007). Patchiness of prescribed burns in dry sclerophyll eucalypt forests in South-eastern Australia. *Forest Ecology and Management* **252**, 24–32.

Pople AR, Phinn SR, Menke N, Grigg GC, Possingham HP and McAlpine C (2007). Spatial patterns of kangaroo density across the South Australian pastoral zone over 26 years:

Aggregation during drought and suggestions of long distance movement. *Journal of Applied Ecology* **44**, 1068–1079.

Salzer D and Salafsky N (2006). Allocating resources between taking action, assessing status, and measuring effectiveness of conservation actions. *Natural Areas Journal* **26**, 310–316.

Shea KP, Amarasekare P, Kareiva P, Mangel M, Moore J, Murdoch WW, Noonburg E, Parma AM, Pascual MA, Possingham HP, Wilcox C and Yu D (1998). Management of populations in conservation, harvesting and control. *Trends in Ecology and Evolution* **13**, 371–375.

Strayer D, Glitzenstein JS, Jones CG, Kolasoi J, Likens GE, McDonnell MJ, Parker GG and Pickett STA (1986). 'Long-term ecological studies: An illustrated account of their design, operation, and importance to ecology'. Occasional Publication of the Institute of Ecosystem Studies, No. 2. Institute of Ecosystem Studies, Millbrook, New York.

Taylor BL and Gerrodette T (1993). The uses of statistical power in conservation biology: The vaquita and northern spotted owl. *Conservation Biology* **7**, 489–500.

Tenhumberg B, Tyre AJ, Shea K and Possingham HP (2004). Linking wild and captive populations to maximize species persistence: Optimal translocation strategies. *Conservation Biology* **18**, 1–11.

Trewhella WJ, Rodriguez-Clark KM, Corp N, Entwistle A, Garrett SRT, Granek E, Lengel KL, Raboude MJ, Reason PF and Sewall BJ (2005). Environmental education as a component of multidisciplinary conservation programs: Lessons from conservation initiatives for critically endangered fruit bats in the western Indian Ocean. *Conservation Biology* **19**, 75–85.

Vaughan H, Whitelaw G, Craig B and Stewart C (2003). Linking ecological science to decision-making: Delivering environmental monitoring information as societal feedback. *Environmental Monitoring and Assessment* **88**, 399–408.

Walters CJ (1986). *Adaptive Management of Renewable Resources*. Blackburn Press, Caldwell, New Jersey, USA.

Whitelaw G, Vaughan H, Craig B and Atkinson D (2003). Establishing the Canadian community monitoring network. *Environmental Monitoring and Assessment* **88**, 409–418.

Wintle BA, Runge MC and Bekessey SA (2010). Allocating monitoring effort in the face of unknown unknowns. *Ecology Letters* **13**, 1325–1337.

Yoccoz NG, Nichols JD and Boulinier T (2001). Monitoring of biological diversity in space and time. *Trends in Ecology and Evolution* **16**, 446–453.

7 BIG-PICTURE ASSESSMENT OF BIODIVERSITY CHANGE: SCALING UP MONITORING WITHOUT SELLING OUT ON SCIENTIFIC RIGOUR

Simon Ferrier

Lesson #1. Avoid reinventing the scientific wheel – build on previous advances in thinking about biodiversity monitoring.

Lesson #2. Allow for diverse forms of monitoring to serve diverse policy, planning and management needs.

Lesson #3. Better define the role and purpose of big-picture biodiversity change assessment relative to other forms of monitoring.

Lesson #4. Get smarter in linking top-down (remote sensing) and bottom-up (field-based) monitoring paradigms.

Lesson #5. Recognise the crucial contribution that modelling can make to achieving this integration.

Lesson #6. View multiple components of biodiversity monitoring, assessment and decision-making within a broader integrated framework.

Lesson #7. Be receptive to the potential value of new observation technologies.

Introduction

Sitting down to write this chapter I find myself looking back over 30-plus years of involvement in various forms of biodiversity assessment, and reflecting on how my perspective on monitoring has evolved during this period. My involvement with biodiversity monitoring began before the term 'biodiversity' was even coined, when I studied the ecology and conservation status of the Rufous Scrub-bird (*Atrichornis rufescens*) for my PhD thesis (Ferrier 1984). As part of this work, I developed a methodology for monitoring change in the abundance of this rare species, based on counts of calling males made along walked transects. This technique was relatively sophisticated in accounting for the effects of season, time of day, weather, topography and habitat on the detectability of individual birds. It was, however, a methodology designed to answer some very specific questions relating to the conservation management of a single species within a set of relatively small geographical areas.

Fast-forwarding to 2011, the types of questions about biodiversity change that I have been asked to help answer in recent times could not be more different from the situation described above. These questions often concern change in biodiversity 'as a whole' – i.e. compositional, structural and functional diversity across all taxa, and across all levels of biological organisation from genes to species to ecosystems. They also typically relate to very large geographical areas – for example the whole Australian continent, or even the entire planet. Answers to such big-picture questions are often difficult to derive through direct field-based monitoring alone, and approaches employed in this work therefore rely heavily on various forms of remote sensing and model-based inference (Ferrier 2011). Most of the lessons I have chosen to discuss in this chapter relate to challenges of big-picture monitoring of biodiversity change – a perspective somewhat different, and therefore hopefully complementary, to that adopted in the other chapters. In a few of these lessons, I also direct particular attention to recent advances in, and future prospects for, more effectively integrating big-picture, or 'top-down', monitoring (based largely on remote sensing) with more localised 'bottom-up' field monitoring – thereby closing the loop with the type of monitoring I cut my teeth on with the Rufous Scrub-bird all those years ago.

Lessons

1. Avoid reinventing the scientific wheel – build on previous advances in thinking about biodiversity monitoring

Numerous papers have already been published in the scientific journal literature promoting best-practice principles for designing and implementing biodiversity (or ecological) monitoring schemes. One of my favourites is a paper by Failing and Gregory (2003) who identified 'ten common mistakes in designing biodiversity indicators for forest policy'. These included:

1. failing to define endpoints;
2. mixing means and ends;
3. ignoring the management context;
4. making lists instead of indicators;
5. avoiding importance weights for indicators;
6. avoiding summary indicators or indices;
7. failing to link indicators to decisions;
8. confusing value judgements with technical judgements;
9. substituting data collection for critical thinking; and
10. oversimplifying: ignoring spatial and temporal trade-offs.

The wisdom and advice conveyed by this paper is as relevant today as it was eight years ago, and many of the principles promoted by Failing and Gregory (2003) have been echoed, and explored in greater depth, in several other recent publications dealing with this general topic (e.g. Lindenmayer and Likens 2010). Hopefully, therefore, these fundamental principles of best practice in biodiversity monitoring are already well understood and well accepted throughout the relevant scientific community. In my opinion, the most significant ongoing challenge now facing this community is to achieve more effective translation, or mainstreaming, of these principles into well-designed and well-resourced monitoring programs, through closer collaboration with relevant agencies and organisations at regional, state, national and global levels.

2. Allow for diverse forms of monitoring to serve diverse policy, planning and management needs

One of the most commonly promoted principles of best-practice is that any given biodiversity monitoring scheme or program should be linked to, and thereby serve, a clearly defined decision-making need. This principle is often illustrated in the scientific literature using examples focused on decisions relating to the application of particular management actions at particular locations to achieve conservation outcomes for particular species. Yet such decisions, as important as they are, constitute just a fraction of the diverse array of real-world policy, planning and management decisions affecting biodiversity outcomes (Ferrier and Wintle 2009). These decisions can vary greatly in: (1) the type of action resulting from the decision (e.g. formulation of environmental policy or legislation; allocation of funding and other resources to high-level environmental programs or regions; implementation of on-ground management); (2) the particular value(s) of biodiversity that the decision is intended to enhance (e.g. conservation of individual species of particular societal concern; conservation of biodiversity as a whole; provision of ecosystem services); and (3) the extent of the spatial domain affected by the decision (e.g. an individual paddock; a catchment management area; a state; the whole continent; the entire planet). Different types of policy, planning and management decisions will often need to be informed by different types of monitoring.

3. Better define the role and purpose of big-picture biodiversity change assessment relative to other forms of monitoring

Effective linking of, and establishment of adaptive feedbacks between, monitoring and decision-making is relatively easy to envisage for decisions involving narrowly defined values (e.g. conservation of particular species), management options and geographical domains. These links become more difficult to envisage, and even harder to implement, as the scope of decision-making broadens. Efforts to assess and report big-picture changes in overall biodiversity at continental and global scales may be, or at least appear to be, particularly disconnected from decision-making. Yet, the same principles of best-practice that have been formulated for more tightly focused monitoring activities (see Lesson #1 above) should be just as relevant at these broader scales – including the desirability of linking monitoring to clearly defined decision-making needs.

Even global-scale monitoring of biodiversity change can, and should, be linked explicitly to decision-making (Scholes *et al.* 2008; Butchart *et al.* 2010; Jones *et al.* 2011). While the types of decisions that need to be informed by monitoring at this scale are very different from those informed by more focused monitoring activities, they are decisions often of considerable consequence for the achievement of biodiversity conservation outcomes. For example, big-picture biodiversity monitoring can play a key role in informing decisions regarding establishment of environmental policy and regulatory mechanisms, and high-level allocation of funding to environmental programs and initiatives. To play this role effectively, such monitoring may need to be well aligned with nationally or globally agreed biodiversity goals and targets (e.g. Australia's National Biodiversity Strategy www.environment.gov.au/biodiversity/strategy/, or the Convention on Biological Diversity's new 2020 Biodiversity Targets www.cbd.int/sp/). Big-picture monitoring should also inform lower-level decisions about where to most cost-effectively direct scarce conservation resources within a given program, by providing fundamental spatially explicit information on rates of change in the state of biodiversity, and in the pressures acting on this state.

4. Get smarter in linking top-down (remote sensing) and bottom-up (field-based) monitoring paradigms

Big-picture biodiversity monitoring – i.e. assessment of change in overall biodiversity across very broad spatial extents – often relies heavily on satellite-borne remote sensing, or other forms of remote mapping. These technologies are used to track changes in the extent and condition of ecosystem or vegetation types, or changes in the distribution of relevant pressures and management responses. However, while the list of structural and functional attributes of ecosystems measurable through remote sensing continues to grow, remote monitoring of change in biodiversity composition remains a major challenge. This is because most biological elements constituting compositional diversity, especially at the genetic and species level, cannot (yet) be detected using satellite-borne remote sensing. They can be observed only through direct field-based observation.

Top-down (remote sensing) and bottom-up (field-based) approaches to biodiversity change assessment clearly have complementary strengths. Field-based monitoring provides direct information on changes in biodiversity that are difficult, if not impossible, to detect in any other way. Remote sensing, on the other hand, provides a rapid and relatively cost-effective means of inferring changes in biodiversity across the vast majority of the planet's surface that falls beyond the reach of field monitoring efforts. Given this high degree of complementarity, it is unfortunate that these two approaches have evolved largely as separate paradigms, pursued within two distinct scientific communities. As I have recently argued elsewhere (Ferrier 2011), much now stands to be gained from forging stronger links between these two paradigms, and thereby enhancing the value of information returned per unit of expenditure on biodiversity change assessment.

5. Recognise the crucial contribution that modelling can make to achieving this integration

One of the most promising means of integrating the respective strengths of top-down and bottom-up approaches to biodiversity monitoring is through modelling (Ferrier 2011). Big-picture biodiversity assessments, including those at continental and global levels, often report changes against long lists of remotely mapped indicators (e.g. Butchart et al. 2010). These typically include state indicators (e.g. extent of different ecosystem types), pressure indicators (e.g. human-population density, ecological footprint), and response indicators (e.g. protected-area coverage). Relating these sorts of remotely mapped indicators to field-based observations of biodiversity change is challenging because, more often than not, the two approaches are actually measuring quite different things (versus measuring the same thing using different techniques and/or units). The remotely mapped indicators essentially estimate change in various factors that are expected to, in combination, play a role in shaping biodiversity change. But, unlike field-based monitoring, these indicators do not typically provide any estimate of the actual change in biodiversity resulting from the interplay of these factors.

Modelling can play a crucial role in bridging this divide. Rather than remotely mapped data simply being used to generate state, pressure and response indicators, these same data can serve as best-available estimates of key factors in an integrated model to predict, or infer, the state of biodiversity itself. This capacity to translate remotely mapped changes in relevant factors into explicit predictions of biodiversity changes expected on the ground, opens the way to using field-based monitoring to evaluate and calibrate predictions, thereby enabling adaptive refinement of the underlying models over time. Such modelling is still very much in its infancy, but experience from recent years suggests that this general approach holds considerable promise. For example, in my own work, we have been experimenting with a suite of modelling techniques for inferring the proportion of species diversity expected to be retained in extensive geographical areas of interest (e.g. bioregions, countries, continents) as a function of multiple remotely generated data sets of the type traditionally employed in deriving state, pressure and response indicators (e.g. Ferrier et al. 2004; Allnutt et al. 2008; Faith et al. 2008; Williams et al. 2010). Once predictions from such models start to be evaluated using field-based observations, they may well exhibit relatively high rates of error. But this process of iterative model evaluation and calibration should allow models to be progressively improved over time, and thereby further strengthen the coupling of top-down and bottom-up monitoring approaches into the future.

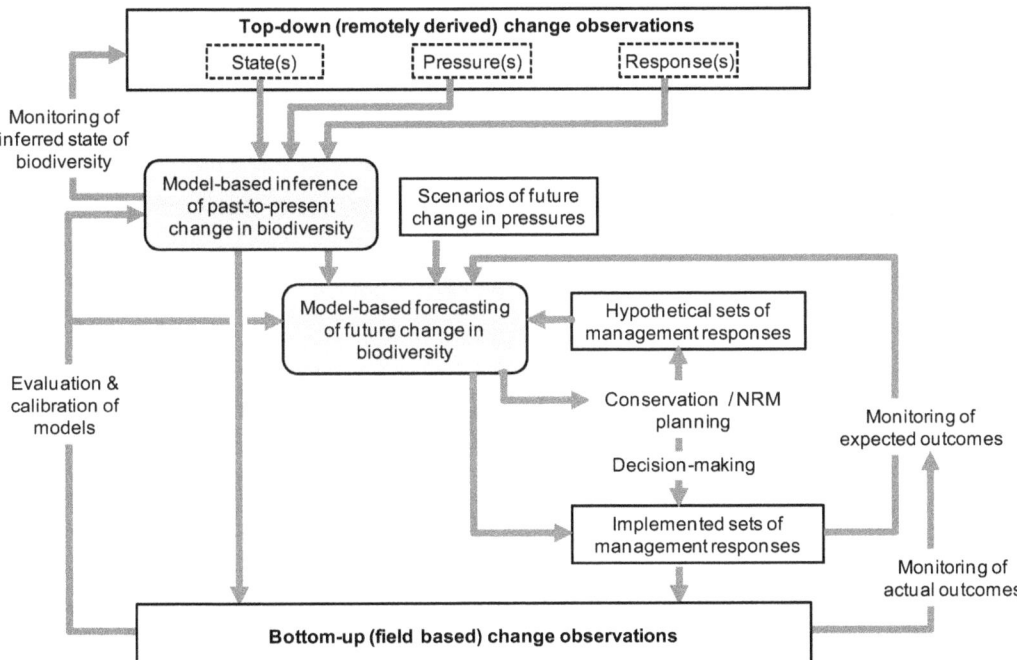

Figure 7.1. Model-based integration of top-down and bottom-up approaches to biodiversity change observation, as part of a broader framework of monitoring, assessment and decision-making.

6. View multiple components of biodiversity monitoring, assessment and decision-making within a broader integrated framework

This use of modelling to link remote-sensing and field-based monitoring programs can, in turn, be viewed as forming part of a broader integrated framework of biodiversity monitoring, assessment and decision-making (see Figure 7.1). Models used to integrate multiple remotely mapped factors to infer the changing state of biodiversity (past to present) also offer an important foundation for predicting, or forecasting, change in biodiversity into the future, under alternative scenarios of drivers and pressures (e.g. land-use change, climate change) and alternative, spatially explicit configurations of management responses (e.g. Ferrier and Drielsma 2010). Such scenario-based forecasts provide, in turn, an objective basis for making decisions about where best to direct scarce conservation resources to maximise expected return on investment.

Modelling plays one more key role in the framework depicted in Figure 7.1. Discussions about biodiversity monitoring, particularly in natural resource management (NRM), often make a strong distinction between monitoring of 'activities' – i.e. amounts of management actions implemented (e.g. kilometres of stock-exclusion fencing) – versus monitoring of 'outcomes' – i.e. actual changes in biodiversity observed to result from these actions (Hobbs, Chapter 5, this volume). Monitoring of outcomes, rather than just activities, is clearly regarded as best practice. However, the impact of implemented actions may often take many years, or even decades, to be manifested in observable changes in the state of biodiversity. Such time-frames do not align well with the relatively short reporting cycles of most NRM processes. Modelling can potentially help to alleviate this situation by offering a third form of monitoring – i.e. monitoring of 'expected outcomes' – that sits between the two extremes of activity and outcome monitoring (Ferrier 2005). Using the same models employed in decision-making (described above) to, in turn, forecast the collective benefit of sets of actions actually implemented on the ground can help provide more rapid feedback on expected return on investment. Under this approach, outcome monitoring (through rigorous field-based observation) still plays a crucial role by enabling evaluation and adaptive refinement of model-based forecasts of biodiversity outcomes.

7. Be receptive to the potential value of new observation technologies

Both top-down and bottom-up approaches to monitoring biodiversity change stand to benefit greatly in coming years from new and emerging observation technologies. Practitioners in biodiversity assessment and monitoring often harbour a healthy wariness or scepticism of new technological developments. This is understandable, given the past tendency for proponents of technologies such as remote sensing and molecular biology to over-sell the potential of these methodologies for biodiversity monitoring applications. However, current advances in both of these fields are now starting to offer benefits that will be increasingly difficult to ignore. From the world of remote sensing, the advent of new satellite-borne LIDaR and hyper-spectral sensors heralds a new era in the detection and mapping of vegetation structure and floristic composition respectively (Wang *et al.* 2010). From the world of molecular biology, fast-throughput environmental metagenomics (or 'ecogenomics') is quickly opening up potential for very rapid field-sampling and sequencing of components of biodiversity previously regarded as intractable from a monitoring perspective. Application of this technology is no longer confined to microbial diversity, with recent demonstrations of its potential for monitoring change in other biological groups, including invertebrates (e.g. Chariton *et al.* 2010).

Conclusion

Diverse forms of biodiversity monitoring are needed to inform policy, planning and management decisions focusing on different values, and involving different types of actions, at different scales. Efforts to assess big-picture changes in overall biodiversity at continental and global scales often rely heavily on satellite-borne remote sensing, or other forms of remote mapping. This top-down approach to biodiversity change observation has evolved largely as a separate paradigm to that of field-based (bottom-up) monitoring. Modelling can play a crucial role in bridging this divide, by translating remote observations of relevant factors into explicit predictions of changes in biodiversity expected on the ground. Field-based monitoring can then be used to evaluate and calibrate these predictions, thereby enabling adaptive refinement of models over time. Such model-based linking of remote-sensing and field-based monitoring programs should form part of a broader integrated framework of biodiversity monitoring, assessment and decision-making

Acknowledgements

Development of the ideas presented here has benefited greatly from interactions with many inspiring conservation scientists and practitioners with whom I have had the privilege of working over the years.

Biography

Simon Ferrier is a Senior Principal Research Scientist with CSIRO Ecosystem Sciences, based in Canberra. Up until three years ago he worked with the NSW Department of Environment,

Climate Change and Water. He has 30 years of experience in researching, developing and applying quantitative spatial analysis and modelling approaches in biodiversity assessment and conservation planning.

References

Allnutt TF, Ferrier S, Manion G, Powell GVN, Ricketts TH, Fisher BL, Harper GJ, Irwin ME, Kremen C, Labat J-N, Lees DC, Pearce TA and Rakotondrainibe F (2008). A method for quantifying biodiversity loss and its application to a 50-year record of deforestation across Madagascar. *Conservation Letters* **1**, 173–181.

Butchart SH, Walpole, M, Collen B, van Strien A, Scharlemann JP, Almond RE, Baillie JE, Bomhard B, Brown C, Bruno J, Carpenter KE, Carr GM, Chanson J, Chenery AM, Csirke J, Davidson NC, Dentener F, Foster M, Galli A, Galloway JN, Genovesi P, Gregory RD, Hockings M, Kapos V, Lamarque JF, Leverington F, Loh J, McGeoch MA, McRae L, Minasyan A, Hernández Morcillo M, Oldfield TE, Pauly D, Quader S, Revenga C, Sauer JR, Skolnik B, Spear D, Stanwell-Smith D, Stuart SN, Symes A, Tierney M, Tyrrell TD, Vié JC and Watson R (2010). Global biodiversity: indicators of recent declines. *Science* **328**, 1164–1168.

Chariton AA, Court LN, Hartley DM, Colloff MJ and Hardy CM (2010). Ecological assessment of estuarine sediments by pyrosequencing eukaryotic ribosomal DNA. *Frontiers in Ecology and the Environment* **8**, 233–238.

Failing L and Gregory R (2003). Ten common mistakes in designing biodiversity indicators for forest policy. *Journal of Environmental Management* **68**, 121–132.

Faith DP, Ferrier S and Williams KJ (2008). Getting biodiversity intactness indices right: ensuring that 'biodiversity' reflects 'diversity'. *Global Change Biology* **14**, 207–217.

Ferrier S (1984). The status of the Rufous Scrub-Bird *Atrichornis rufescens*: habitat, geographical variation and abundance. PhD Thesis. University of New England, Armidale, Australia.

Ferrier S (2005). 'An integrated approach to addressing terrestrial biodiversity in NRM investment planning and evaluation across state, catchment and property scales'. A submission to the NSW Natural Resources Commission. NSW Department of Environment and Conservation, Sydney. <http://www.nrc.nsw.gov.au/content/documents/Submission%20-%20ST%20-%20DEC%20Biodiversity.pdf>

Ferrier S (2011). Extracting more value from biodiversity change observations through integrated modeling. *BioScience* **61**, 96–97.

Ferrier S and Drielsma M (2010). Synthesis of pattern and process in biodiversity conservation assessment: a flexible whole-landscape modelling framework. *Diversity and Distributions* **16**, 386–402.

Ferrier S and Wintle B (2009). Quantitative approaches to spatial conservation prioritisation: matching the solution to the need. In *Spatial Conservation Prioritization: Quantitative Methods and Computational Tools*. (Eds A Moilanen, H Possingham and K Wilson) pp. 1–15. Oxford University Press, Oxford.

Ferrier S, Powell GVN, Richardson KS, Manion G, Overton JM, Allnutt TF, Cameron SE, Mantle K, Burgess ND, Faith DP, Lamoreux JF, Kier G, Hijmans RJ, Funk VA, Cassis GA, Fisher BL, Flemons P, Lees D, Lovett JC and van Rompaey RSAR (2004). Mapping more of terrestrial biodiversity for global conservation assessment. *BioScience* **54**, 1101–1109.

Jones JPG, Collen B, Atkinson G, Baxter PW, Bubb P, Illian JB, Katzner TE, Keane A, Loh J, McDonald-Madden E, Nicholson E, Pereira HM, Possingham HP, Pullin AS, Rodrigues

AS, Ruiz-Gutierrez V, Sommerville M and Milner-Gulland EJ (2011). The why, what, and how of global biodiversity indicators beyond the 2010 target. *Conservation Biology* **25**, 450–457.

Lindenmayer DB and Likens GE (2010) The science and application of ecological monitoring. *Biological Conservation* **143**, 1317–1328.

Scholes RJ, Mace GM, Turner W, Geller GN, Jürgens N, Larigauderie A, Muchoney D, Walther BA and Mooney HA (2008). Toward a global biodiversity observing system. *Science* **321**, 1044–1045.

Wang K, Franklin SE, Guo XL and Cattet M (2010). Remote sensing of ecology, biodiversity and conservation: a review from the perspective of remote sensing specialists. *Sensors* **10**, 9647–9667.

Williams KJ, Ferrier S, Rosauer D, Yeates D, Manion G, Harwood T, Stein J, Faith DP, Laity T and Whalen A (2010). 'Harnessing continent-wide biodiversity datasets for prioritising national conservation investment'. A report to the Department of the Environment, Water, Heritage and the Arts. CSIRO, Canberra.

8 AN ENDPOINT HIERARCHY AND PROCESS CONTROL CHARTS FOR ECOLOGICAL MONITORING

Mark Burgman, Kim Lowell, Peter Woodgate, Simon Jones, Gary Richards and Prue Addison

Lesson #1. Monitoring variables should have a social mandate.

Lesson #2. Monitoring variables should be selected from an endpoint hierarchy to ensure they satisfy operational and scientific objectives.

Lesson #3. Monitoring data should be linked directly and explicitly to a decision context.

Lesson #4. Null hypothesis tests are the wrong way to use monitoring data to make decisions.

Lesson #5. Control charts should be developed and tuned to the relevant decision context.

Lesson #6. Control charts may be aggregated to contribute to regional report cards that satisfy social and political, as well as scientific, goals.

Introduction

An environmental monitoring framework should address scientific objectives and local needs, and be scalable so that it provides relevant information on the implementation of government policies. More specifically, monitoring frameworks should provide:

- Unambiguous, sensitive feedback on ecological attributes of interest to scientists and land managers (sizes of threatened populations, extent and condition of vegetation, water quality and so on);
- Information for prioritising government and private investments to improve catchment management and protect remnant vegetation; and
- Information to encourage community engagement in conservation.

The history of ecology is littered with monitoring programs that either fail to satisfy scientific objectives or are jettisoned as soon as budgets contract. The purpose of this chapter is to describe a monitoring framework that employs a number of techniques that have been successful in a variety of fields. These include: the use of an endpoint hierarchy for identifying appropriate measurement systems, a technique that has been used by ecotoxicologists to address a raft of scientific, social and operational constraints since the 1990s (Suter 1993), and the use of Process Control techniques as a means of displaying monitoring information in simple, practical and scientifically credible ways which were developed largely for manufacturing applications, mainly between the 1930s and the 1970s (Montgomery 2001). The use of these techniques within a complete monitoring framework for environmental measurement has, at best, been limited (but see Anderson and Thompson 2004; Mesnil and Petitgas 2009). Finally we re-visit the utility of regionalised report cards (Wentworth Group of Concerned Scientists 2008) for displaying the results of environmental monitoring and trend analyses and documenting progress towards environmental goals using monitoring data by using a hypothetical case study for monitoring native vegetation.

Lessons

1. Monitoring variables should have a social mandate

The priority given to any monitoring activity can be determined by the degree to which the need for action is socially and politically mandated, the likelihood that an unacceptable change will occur, the plausibility of implementing effective remedial action if the projected change occurs, and costs and logistical constraints of monitoring.

The social relevance of environmental management activities, including monitoring, can be determined through stakeholder consultation. This step matters because monitoring protocols require a social mandate, although exploring techniques for this are beyond the scope of this chapter. Monitoring activities that fail to address social priorities are doomed to fail in political arenas, irrespective of their scientific or practical merit. Of course, monitoring priorities should be further refined and reviewed by technical experts to ensure that they are operationally feasible and ecologically relevant.

The prioritisation and consultation process should result in a set of endpoints that enable land and resource managers to assign priority to a given issue and then to determine what management actions and monitoring measurements are required. This in turn allows managers to provide assurance to stakeholders that primary concerns are being addressed in a systematic way.

2. Monitoring variables should be selected from an endpoint hierarchy to ensure they satisfy operational and scientific objectives

Suter (1993) described an endpoint hierarchy for identifying an appropriate measurement system to define, monitor and manage environmental change. The hierarchy defines three types of endpoints as an expression of the values that it is desired to protect based on a raft of scientific, social and operational constraints. Specifics of these three types of endpoints can be developed for each issue or ecosystem of interest.

1. *Management Goals* embody broad policy objectives and carry a clear social mandate. For environmental systems, management goals may be targeted at an ecosystem service or attribute (Chee 2004). For example, they may include the 'maintenance of natural vegetation' or the 'protection of threatened species'.
2. *Assessment Endpoints* translate the management goals through a conceptual model into a more pragmatic context. They are more specific than the management goals but faithfully reflect the social mandate. An example of an assessment endpoint relevant to maintaining natural vegetation would be 'the extent of vegetation communities' or the 'condition of vegetation' within a relevant land-use class.
3. *Measurement Endpoints* address the actual measurement of items of interest, and address issues of operational feasibility, sample size and logistics. For instance, to monitor the extent and condition of natural vegetation, a manager may record the area of each vegeta-

tion type in the area of interest annually, using remote sensing and aerial photography, and require field re-measurement of randomly located monitoring sites within a subset of the vegetation types.

Suter (1993) suggested measurement endpoints should be biologically relevant, important to society, unambiguously defined, operationally feasible, predictable and measurable, and susceptible to the stresses affecting ecosystem processes. Moreover, if measurement endpoints are to provide credible evidence that management goals have been attained, they must also satisfy the expectations of policy makers, politicians, and others (Crawford-Brown 1999; Burgman 2005).

Measurement endpoints are essentially indicators used to infer success or failure of management goals. Most measurement endpoints are local in nature, but should be selected considering the need to aggregate to a state or national level (Parkes *et al.* 2003; Stone *et al.* 2003). Measurement endpoints need to be measurable with sufficient statistical integrity to detect changes in targeted resources reliably, and to detect critical changes that are unacceptable for social, economic or ecological reasons (Munkittrick *et al.* 2009).

3. Monitoring data should be linked directly and explicitly to a decision context

Often it is difficult to discern exactly why a monitoring program was established, or how managers are expected to react to monitoring data. Monitoring data are collected to inform managers about the performance of a system. These data should provide guidance on appropriate interventions and timely indications of sudden changes, and of systems that are drifting into unacceptable states. Ideally, they make up a sound, simple, operationally feasible quality assurance process. That is, they support decision-making. Unless the decision context is clearly defined, it is impossible to evaluate whether a monitoring program is appropriately designed and sufficiently sensitive to detect important changes.

Alternative specifications for the monitoring system may be aimed at:

1. *Calibration*; assigning the set of monitoring attributes and linking them to changes that may be unacceptable for social or scientific reasons. This is the single most important task, and may be implemented using the endpoint hierarchy outlined above.
2. *Validation*; systematic auditing to confirm that the sets of monitoring attributes meet predetermined standards and statements of fitness-for-purpose.
3. *Continuous improvement*; monitoring programs must have a feedback mechanism to enable changes in monitoring activities and investments to better monitor the outcomes of environmental management. That is, the monitoring program itself should improve over time, and the expectation for change should be built into all monitoring programs to facilitate the adoption of improved techniques, measurement protocols and field skills.

4. Null hypothesis tests are the wrong way to use monitoring data to make decisions

Null hypothesis tests have long been used as a means of interpreting evidence. Unfortunately, they reflect only the Type I error rate, the chance of acting unnecessarily to protect the environment. Environmental scientists care about both the Type I error rate, and the Type II error rate, the chance of failing to act when it was in fact warranted. Yet many scientists believe that large p-values in tests of the significance of a trend or effect in a monitoring program mean that there is no effect and no need to act. Blindness to Type II error rates leads to routine and substantial misinterpretation of environmental evidence, even in peer-reviewed scientific articles (Fidler *et al.* 2006; Walshe *et al.* 2007), and consequently to unintended, avoidable, unacceptable environmental damage.

5. Control charts should be developed and tuned to the relevant decision context

To avoid mistakes, the results of monitoring need to be conveyed in simple statements and graphics. Presentation of the observations together with statements of the implications should be meaningful, unambiguous and relevant to the issues that drove the need for monitoring (clearly identified through the endpoint hierarchy). Process control techniques are ideally suited to these tasks.

Developed originally for manufacturing applications (Montgomery 2001; Burgman 2005), control charts display monitoring information in intuitive forms. They are designed to encourage understanding of the system under observation and can provide an early warning signal of a system that may be going 'out of control', allowing managers to take appropriate corrective actions, ranging from acquiring additional field information, to making wholesale changes to strategies.

Charts may be constructed from typical ecological monitoring data to clearly illustrate temporal trends in variables such as the mean or variation in vegetation cover in relation to the 'control limits', which represent unacceptable change. Control limits indicate specific management responses and may accommodate measurement error, natural variation and anthropogenic impacts such as fire, land clearance and revegetation. The limits may be 'tuned' over time as evidence accumulates for false positive responses (taking action where it was unwarranted) and false negatives (failing to act when it was necessary).

Setting control limits for control charts requires careful consideration (Montgomery 2001; Anderson and Thompson 2004), as these must allow for the range of responses to disturbance and successional processes of ecological systems. The nature of disturbance in Australia makes it difficult to define single meaningful benchmarks narrowly (McCarthy *et al.* 2004; see Gibbons *et al.* 2008). In a context of natural disturbance in a landscape, the desired properties of a system will change over time, so that system condition should be interpreted against the background of natural variation (e.g. Richards *et al.* 1999; McCarthy *et al.* 2004). Moreover, benchmarks in a monitoring system must relate to economic and social outcomes as well as environmental objectives if they are to satisfy policy objectives related to management goals.

Many control chart types may be developed. They can be tuned to be sensitive to the kinds of data (e.g. univariate or multivariate; see, for example, ANZECC/ARMCANZ 2000) and the kinds of changes in measurement endpoints that may reflect important changes in underlying processes (see Montgomery 2001; Anderson and Thompson 2004; Caulcutt 2004; Mesnil and Petitgas 2009). Finally, triggering management actions by exceeding a threshold is just one of many options. More sensitive decision rules may be based on the frequency, or the spatial or temporal pattern, of particular attributes (Burgman 2005).

6. Control charts may be aggregated to contribute to regional report cards that satisfy social and political, as well as scientific, goals

The development of a monitoring system is generally a response to one or more policy objectives encapsulated in management goals (e.g. 'maintain vegetation condition'). As noted above, such objectives should be the product of social and political engagement and stakeholder consultation.

Having identified the global objective(s), managers must specify the intensity of effort (cost and resources) as well as the assessment endpoints and the measurement endpoints, selecting appropriate mapping and measuring technologies. This may be an iterative process, refined to

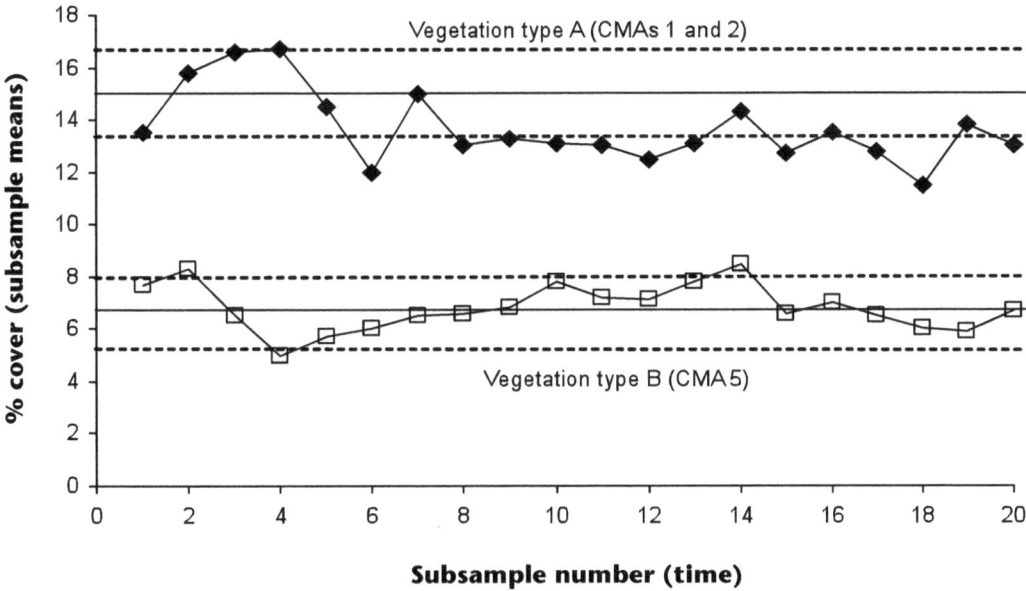

Figure 8.1. Hypothetical report card for remnant vegetation in agricultural landscapes. The dashed horizontal lines are the control limits.

ensure the vegetation mapping and monitoring program is feasible and cost-effective. All maps require some form of calibration to ensure that mapped areas are attributed consistently and with an acceptable level of accuracy. The mix of field and remotely sensed data will be specific to each measurement endpoint and management goal.

A 'report card' (Wentworth Group of Concerned Scientists 2008) may be used to display the results of the monitoring and any trend analyses (see Figure 8.1). The report card should include the issue(s) that prompted the monitoring, the management goal(s) and the level of reporting. This should include, where relevant, the presence or absence of change, the cause, the trend against the reference point or benchmark and a statement of implications.

Conclusions

The framework for monitoring can be developed by applying the following protocol:

1. Set management goals after consideration of the issues and aspirations identified through stakeholder consultation;
2. Link broad policy objectives to assessment endpoints and to precise measurement endpoints;
3. Provide information to allow users to assess the fitness-for-purpose of the data (e.g. spatial scale, mapping certainty);
4. Use causal models to anticipate causes of change that provide a basis for constructing and tuning control charts and to report changes in condition;
5. Provide environmental goals, benchmarks and/or a baseline against which the condition can be assessed, and control chart limits be specified, to display monitoring data and specify response actions; and
6. Use a 'report card' to relate policy objectives to assessment and measurement endpoints and ultimately to a quantitative assessment of condition, trends in condition and statements about the management actions and information required.

Biographies

Mark Burgman is Director of the Australian Centre of Excellence for Risk Analysis and the Adrienne Clarke Chair of Botany in the School of Botany at the University of Melbourne. He works on ecological modelling, conservation biology and risk assessment. He received a BSc from the University of New South Wales, an MSc from Macquarie University, Sydney, and a PhD from the State University of New York. He worked as a consultant ecologist and research scientist in Australia, the United States and Switzerland during the 1980s before joining the University of Melbourne in 1990.

Kim Lowell has 20+ years of experience working with geospatial technologies with emphasis on geographic information systems (GISs) and remote sensing for natural resource management. In the last 5 years he has held a dual position as Principal Scientist in Spatial Landscape Modelling with the Department of Primary Industries Victoria, and as a Professorial Fellow at the University of Melbourne in the Cooperative Research Centre for Spatial Information. His primary research interest is the quality of spatial natural resource information and how this impacts natural resource management and real world decision-making.

Peter Woodgate was appointed CEO of the Cooperative Research Centre for Spatial Information (CRCSI) in June 2003. Peter has over 20 years' experience in a range of disciplines spanning life sciences, engineering, business, public policy and administration. Peter is currently the Deputy Chair of the Cooperative Research Centres Association of Australia, a member of the Executive Committee of the International Society of Digital Earth, Director of the Terrestrial Ecosystems Research Network, member of the Sustainability Council of the Victorian Association of Forest Industries and a foundation member and Chairman of the Research Committee of UNESCO's Mornington Peninsula and Westernport Biosphere Reserve.

Simon Jones works in remote sensing and resource mapping, with a particular focus on issues of scale, data uncertainty and the processes of validating and understanding remotely sensed imagery. He previously worked for the Joint Research Centre of the European Commission (Global Vegetation Monitoring Unit) and contributed to mapping the remnant native vegetation of Victoria and New South Wales. He is currently a research leader in large federally funded research initiatives on spatial data interpretation and analysis. He is a foundation member and former director of the (Surveying and) Spatial Sciences Institute, Australia, and is a Professor in Mathematical and Geospatial Sciences at RMIT University.

Gary Richards is a Principal Advisor (International Forest Monitoring) for the Department of Climate Change and Energy Efficiency. Previously, he led the development of Australia's National Carbon Accounting System which pioneered the combination of remotely sensed data, ground data and carbon cycle models for national emissions reporting. Gary is an Adjunct Professor at the ANU Fenner School of Environment and Society and is also co-chair of the Clinton Climate Initiative, Carbon Measurement Collaborative, participates in the work of the IPCC and Chairs the Task Force of the intergovernmental Group on Earth Observations Global Forest Observations Initiative.

Prue Addison is undertaking her PhD research on the applicability of novel statistical methods (such as control charts) for analysing trends in long-term marine monitoring data sets, such as those collected within Marine Protected Areas. Her research is supervised by Professor Mark

Burgman and Dr Jan Carey at the University of Melbourne and is in collaboration with Parks Victoria and the Joint Nature Conservation Committee in the UK. Prue's research is focused on assessing the utility of these statistical methods for detecting important changes in marine biological communities.

References

Anderson MJ and Thompson AA (2004). Multivariate control charts for ecological and environmental monitoring. *Ecological Applications* **14**, 1921–1935.

ANZECC/ARMCANZ (2000). 'Australian guidelines for water quality monitoring and reporting'. Australian and New Zealand Environment and Conservation Council, Agriculture and Resource Management Council of Australia and New Zealand. Australian Government Publishing Service, Canberra.

Burgman MA (2005). *Risks and Decisions for Conservation and Environmental Management.* Cambridge University Press, Cambridge, UK.

Caulcutt R (2004). Control charts in practice. *Significance* **1**, 81–84.

Chee YE (2004). Economic assessment and valuation of ecosystem services. *Biological Conservation* **120**, 549–565.

Crawford-Brown D (1999). *Risk-based Environmental Decisions: Method and Culture.* Kluwer Academic, Boston.

Fidler F, Burgman MA, Cumming G, Buttrose R and Thomason N (2006). Impact of criticism of null hypothesis significance testing on statistical reporting practices in conservation biology. *Conservation Biology* **20**, 1539–1544.

Gibbons P, Briggs SV, Ayers DA, Doyle S, Seddon J, McElhinny C, Jones N, Sims R and Doody JS (2008). Rapidly quantifying reference conditions in modified landscapes. *Biological Conservation* **141**, 2483–2493.

McCarthy M, Parris K, van der Ree R, McDonnell M, Burgman M, Williams N, McLean N, Harper M, Meyer R, Hahs A and Coates T (2004). The habitat hectares approach to vegetation assessment: an evaluation and suggestions for improvement. *Ecological Management & Restoration* **5**, 24–27.

Mesnil B and Petitgas P (2009). Detection of changes in time-series of indicators using CUSUM control charts. *Aquatic Living Resources* **22**, 187–192.

Montgomery DC (2001). *Introduction to Statistical Quality Control.* Wiley, New York.

Munkittrick KR, Arens CJ, Lowell RB and Kaminski GP (2009). A review of potential methods of determining critical effect size for designing environmental monitoring programs. *Environmental Toxicology and Chemistry* **28**, 1361–1371.

Parkes D, Newell G and Cheal D (2003). Assessing the quality of native vegetation: the Habitat Hectares approach. *Ecological Management & Restoration* **4(S1)**, S29–S38.

Richards SA, Possingham HP and Tizard J (1999). Optimal fire management for maintaining community diversity. *Ecological Applications* **9**, 880–892.

Stone C, Wardlaw T, Floyd R, Carnegie A, Wylie R and deLittle D (2003). Harmonisation of methods for the assessment and reporting of forest health in Australia – a starting point. *Australian Forestry* **66**, 233–246.

Suter GW (1993). *Ecological Risk Assessment.* Lewis Publishers, Michigan.

Walshe T, Wintle B, Fidler F and Burgman M (2007). Use of confidence intervals to demonstrate performance against forest management standards. *Forest Ecology and Management* **247**, 237–245.

Wentworth Group of Concerned Scientists (2008). 'Accounting for nature. A model for building the national environmental accounts of Australia'. Wentworth Group of Concerned Scientists, Sydney.

9 LESSONS FROM ENVIRONMENT ACCOUNTING FOR IMPROVING BIODIVERSITY MONITORING

Michael Vardon

Lesson #1. Build on the past.

Lesson #2. Must have sound institutional arrangements and legal basis.

Lesson #3. Learn by doing and accept what you have.

Lesson #4. Regular and ongoing beats infrequent and *ad hoc*.

Lesson #5. Build capacity of teams to deliver data.

Lesson #6. Integration of biodiversity data with other data is critical.

Lesson #7. Determine what to measure and how to measure it.

Lesson #8. Decide how much is enough for effective monitoring.

Lesson #9. Be able to access and interpret data.

Lesson #10. Define the questions and maintain flexibility.

Introduction

Environmental accounting is a relatively new discipline. It has evolved nationally and internationally on a range of fronts and can take a variety of forms, but usually has the aim of representing the interactions within and between the physical world (or the environment) with the activities of people (or the economy).

Internationally environmental accounting has been formalised in the System of Environmental-Economic Accounting (SEEA) (UN 1993; UN 2003). Smith (2007) provides an introduction to the SEEA and its development. Some key dimensions of data quality from the paradigm of national statistical agencies, of which the Australian Bureau of Statistics (ABS) is one, can be found in Statistics Canada (2002).

Ecosystems and biodiversity are recognised in the SEEA, but scant attention is given to them and to date there are no 'official' accounts for ecosystems and biodiversity in the world, although there has been some progress made in academia and in international agencies in Europe (see Weber 2007). An updated SEEA should provide more clarity on ecosystem and biodiversity accounts at a theoretical level and there are some efforts nationally and internationally to produce ecosystem accounts. This is encouraging.

Even more encouraging is outcome 3.3.1 of the *Biodiversity Conservation Strategy 2010–2030*, which is:

> 'An increased representation of biodiversity and ecosystem services and goods within national accounts.'

The specific mention of ecosystem services and goods and national accounts in the *Biodiversity Conservation Strategy* provides an opportunity to examine the development and implementation of environmental accounts and what lessons may be relevant to the monitoring of biodiversity and the construction of ecosystem or biodiversity accounts. Environmental accounts are also relevant in the context of analysing the efforts made to protect the environment using policies or laws, and in particular, spending on conservation programs.

Two Australian National Audit Office Reports highlight some of the deficiencies in linking government spending to the achievement of desired outcomes in biodiversity conservation. In one report, $1.3 billion was identified as being spent on programs addressing Australia's threatened species and ecological communities, but it was not clear what was achieved (ANAO 2007). This point was reiterated again in 2008 in respect of the Natural Heritage Trust:

> 'There is little evidence as yet that the programs are adequately achieving the anticipated national outcomes.' (ANAO 2008)

Environmental accounts can link spending on environmental protection (or environmental protection expenditure) to changes in the land use and land cover, possibly to changes in the condition of ecological communities, and potentially to species distribution and abundance.

The building of a comprehensive suite of environmental accounts will not be easy and will require institutionalisation of the collection and management of macro-level data on the environment. Such institutionalisation has been achieved with economic statistics in the form of a System of National Accounts. To quote from the 1984 press release announcing the award of a Nobel Prize for Economic Sciences to Richard Stone for his work on national accounts:

> 'Economic reality – such as it appears during a given period (a month, a quarter or a year) – may be seen as numberless (billions upon billions) transactions between purchasers and sellers. For the sake of survey and analysis of an endlessly detailed and complicated mass of transactions for a nation as an economic unit, it is necessary to find methods for systematic summarizing and aggregating of a reality which, on the micro level, is endlessly complicated. A system for national accounts is a method of achieving simplification and an overview.' (Royal Swedish Academy of Sciences 1984)

A similar type of macro-level summary system, that is, one that aggregates the micro-level interactions within the environment and between the environment and the economy, is the goal of environmental accounts. While the environmental system is more complicated than the economic system, aggregated, regular and reliable sets of accounts should assist government and others to construct a simplified overview of the economy and environment and to therefore support better decisions about the environment.

How to summarise data on the conservation of biodiversity and how to place these data into a system of environmental accounts is still an open question and is not the central focus of this book. However, it is a question we should all consider on both theoretical and practical levels.

At the practical level, a large and ongoing public investment will be needed in the 'machinery' used to regularly collect data and publish environmental accounts. Contrast the large resources needed for this task with the probable level of resources available and the manner in which the resources are typically made available (i.e. a large number of small short-term grants, mostly to researchers).

The rest of this chapter is devoted to outlining the successes and the lessons learned from the development of environment accounts in Australia and overseas. The successes are presented below while the lessons, which mention the failures, also contain ideas for 'the way forward'. The chapter ends with a brief conclusion.

Successes – water accounts

Water accounts are arguably the most advanced type of the various environmental accounts and have been produced in a variety of forms within Australia (e.g. ABS 2000; BoM 2009) and internationally (UN 2009; Vardon *et al.* 2012). They can be used for a range of purposes including: identifying the industries and areas consuming the most water; predicting future patterns of water use; and the impact of reduced water availability on the economy (Vardon *et al.* 2007).

1. Build on the past

Water accounting builds on a long history of data collection and analysis in both science and economics. Hydrologists have made water balances for decades, while economic accounts have been compiled since the 1950s. So while water accounting is a relatively new way of combining physical and economic information on water, much of the basic data were in existence and date back to 1965 in Australia (see Vardon *et al.* 2007 for a summary).

The hydrologic and economic worlds were brought together internationally in the System of Environment and Economic Accounts for Water (SEEA-Water; UN 2007) and the subsequent International Recommendations for Water Statistics (UN 2010). Together these provide clear definitions of the data items related to water, how to arrange them to make them suitable for analysis as well as some practical guidance on methods of collection. In addition, the SEEA-Water defines 12 standard tables for the presentation of data. The inclusion of standard tables means that regular users of data do not have to wrestle with the issue of different presentations on the same information. This also means that tables developed in two areas or countries can be directly compared.

The National Water Initiative called for water accounts and other data to support better management of Australia's water. The ABS first published water accounts in 2000 (ABS 2000) and the most recent publication of the Australian Water Account includes the equivalent of four of the SEEA-Water tables – the physical and monetary supply and use tables (see ABS 2010*a*). Data from the Bureau of Meteorology (BoM) in the Pilot National Water Account (BoM 2009) can be used to populate the asset account (Vardon *et al.* in press). In developing these accounting programs, both the ABS and BoM drew on their considerable experience with collecting and publishing data.

2. Must have sound institutional arrangements and legal basis

An important factor in the development of water accounts in Australia was the existence and introduction of legislation that gave a clear mandate and legal framework for the collection of information and the development of the accounts. In the case of the ABS, these are provided by the *Census and Statistics Act 1905* and the *Australian Bureau of Statistics Act 1975* and, for the BoM, the *Water Act 2007.* Two features of the ABS are of special importance: (1) the confidentiality of data provided by individuals and business; and (2) there is no ministerial sign-off on the publication of data.

Internationally, cooperation between agencies and well-defined legal and institutional frameworks are identified as important factors in the success of water accounts (UN 2009). This will almost certainly be the case for monitoring of biodiversity conservation in Australia and elsewhere.

The role of research agencies in monitoring is an issue for debate. While they clearly have a role in the development and trialling of methods and frameworks, an ongoing role in monitoring is less certain. Monitoring is not usually perceived as research and it is seldom glamorous – it is routine. The skills, institutions and institutional arrangements required for routine and ongoing biodiversity monitoring programs differ from those required for research science.

Australia's national institutions are not designed for large-scale environmental monitoring and biodiversity is no exception. Biodiversity monitoring has been carried out by the Commonwealth Scientific and Industrial Research Organisation (CSIRO), other Commonwealth, state and territory government agencies, academic institutions and non-government organisations (NGOs). Most of this monitoring is at fine scales (although there are notable exceptions such as the Birds Australia Atlas of Australian Birds). Broad-scale national initiatives, such as the National Land and Water Resources Audit, have come and gone.

To make significant advances, an agency must take the lead. The Department of Sustainability, Environment, Water, Population and Communities is currently developing a National Plan for Environmental Information. The plan will be developed over a number of years and it is hoped this will provide a clear outline of the institutional roles and identify a lead agency (as has been done for water, with BoM as the lead agency).

A key aspect of environmental monitoring is the role of the states and territories, which have most of the legal powers relating to natural resources and fund the agencies involved in the management and use of natural resources. Cooperation with the states and territories is essential and a fundamental need is the adoption and use of consistent definitions and compatible methodologies by various jurisdictions.

3. Learn by doing and accept what you have

There are a number of proverbs and sayings that have the sentiment 'the best way to finish something is to start it'. In environmental statistics and accounts there is usually great angst about the availability and quality of data needed to support particular accounts. This is exacerbated because there is usually little prospect that the resources needed to completely plug these deficiencies will be found. This can result in paralysis or, even worse, provide an excuse to do nothing.

We must accept that the data available are not perfect, that the resources available are not sufficient and that, even if extra data are collected, they too will have various imperfections. While this offends our scientific desire for near-perfect answers, we must recognise that major decisions are often based on imperfect data. For example, decisions on the management of the economy, including interest rates, are based on imperfect data from the (economic) System of National Accounts. Many decisions are made with very thin, contentious information and even no data at all. Partial information is better than no information for decision-making. The qualification is that the partial information is not misleading.

While accepting that answers obtained will always be approximate is a first step, the next step is to recognise that data will improve over time with practical experience, theoretical advances and technological innovation. For example, the water supply and use tables produced by the ABS provide estimates in respect of more than 2 million employing businesses and nearly 9 million households, plus government and the not-for-profit entities. The scale of this is enormous and is accomplished using various collection methods, including stratified

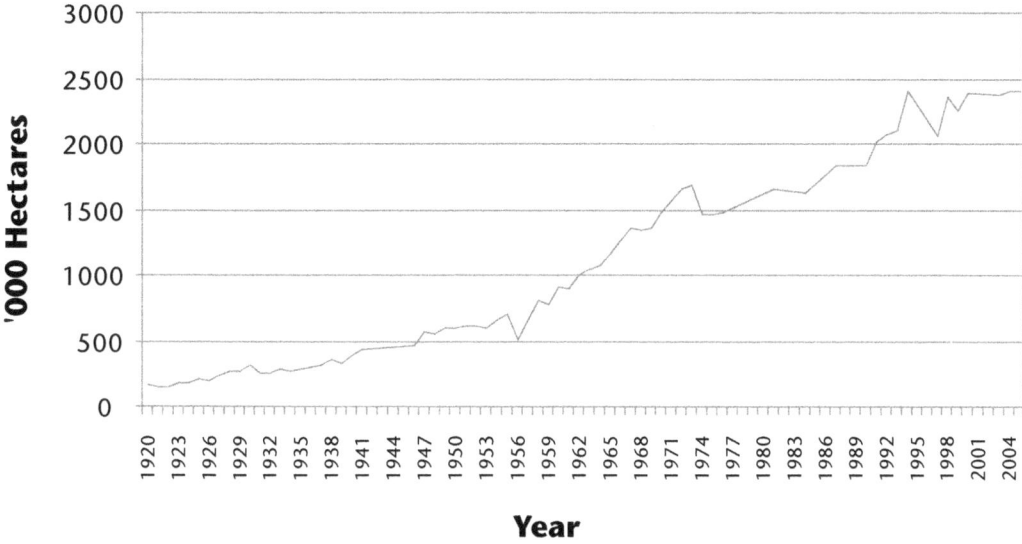

Figure 9.1. Area irrigated in Australia 1920–2005.

random sampling. The sampled fraction is small but has increased a little over time. However, large gains have been made by better targeting of the sample, made possible by improved knowledge of the patterns of water supply and use in industries, as well as the use of improved statistical processing systems and greater use of administrative data collected by the Commonwealth, states and territories.

The timeliness of data and their fitness-for-purpose are key considerations. These can be illustrated by way of example. There is no point getting a flood warning three hours after the flood arrives. If it is clear there is likely to be a peak in waters leading to flooding, then a warning should be issued as soon as possible. In the end it does not matter if the flood peaks at 5.6 or 6.1 metres or at 10.55 or 11.27 a.m. The accuracy with which the peak is correctly predicted is not central to decision-making. A decision to evacuate a town, for example, requires timely rather than 100 per cent accurate data.

4. Regular and ongoing beats infrequent and ad hoc

It is not the first time that is important: it is the second or third, or better still, the 20th or 100th time that data are produced that is important. The bald fact that in 2006 Australia had about 2.5 million ha of irrigated land, is made far more relevant when this snippet of data is placed in a time series spanning nearly 90 years (see Figure 9.1). The world of environmental monitoring is the world of the case study and one-off exercises. A particular achievement of the ABS water account is that it has been produced four times, and will now be produced annually (ABS 2010*b*). This has provided data users, which includes academics and government agencies, with a regular and reliable source of data that they can use for various types of analyses (e.g. predicting the impact of reduced water availability on economic activity).

5. Build capacity of teams to deliver regular data

Producing regular results is important for capacity building. Developing the capacity of teams of people to deliver regular data is a challenge. Funding commitments are often short term and career prospects may be more limited. Finding the right mix of people and skills, providing the necessary training and retaining staff are ongoing challenges for the ABS in the area of environmental accounts.

The move to an annual program of water accounts at the ABS has meant that greater attention can be given to staff retention and training. The previous model involving production of the accounts at 4-year intervals, entailed the creation of process and systems each time and staffing-up to around 12 staff and then down to only a handful. The result of this was that there was little overlap in the teams producing successive accounts, and the systems and processes developed were not maintained, let alone improved. Now a mix of junior and senior staff can be deployed and, because of improvements to basic data sources within and outside the ABS, standard data processing systems can be used and can be improved over time.

Solutions

Improving biodiversity conservation will require many people to work more effectively together. The main issue constraining biodiversity conservation is not the poor understanding of species and their interactions with each other, but on the interactions of biodiversity with human activities, which may be positive or negative. Some will dispute this, but a key part of the way forward is recognising that scientists working alone will not solve the problem of biodiversity conservation. As a society we must be more accountable for the actions that degrade biodiversity as well as realistically assessing the success of the programs or policies introduced specifically to protect biodiversity.

To date most of the focus on monitoring has been on understanding biodiversity at small scales: what species do we have, how many of each species do we have, and how do species interact and respond to natural and anthropogenic change? These are all important questions and there are very good examples of responses to these questions working at fine scales. However, monitoring at broad scales is sadly neglected. There has been little integration of scales, and almost no assessment of the relative cost and benefits of different management options.

To improve the situation the following issues will need to be addressed:

- Institutional arrangements for environmental monitoring
- Relations between biodiversity data and other data
- Determining the metrics for biodiversity measurement
- Deciding how much is enough data for effective monitoring
- Access to data
- Interpretation of data
- The right questions and maintaining flexibility.

Lessons

6. Integration of biodiversity data with other data is critical

Perhaps the greatest challenge for data collectors is how to integrate new data with various other data. This is one of the purposes of environmental accounts. Relating biodiversity monitoring (e.g. on the distribution and abundance of species) to other initiatives is a critical need. For example, the establishment of national parks, introduction of laws or policies to protect species, the implementation of weed or feral animal control programs, spending on environmental protection and biodiversity management in particular, should all be linked with outcomes. To date linking expenditure to the outcomes has not occurred and as a result we have little idea about the cost-effectiveness of the different interventions that can be made by governments, the private sector and community groups.

We should be able to answer questions like: Is 'Bushtender' (Victorian Government 2011) or other forms of government payment to private landowners to manage native vegetation, more or less efficient than public ownership of land (e.g. national parks) for biodiversity conservation? Only an integrated environmental and economic data system can answer this.

The spatial linking of land cover change to economic activities (e.g. agriculture) and government initiatives (subsidies, taxes, regulation, price reform) via a series of environmental accounts (specifically land and environmental protection expenditure accounts) would enable the efficiency of different government or private sector initiatives to be assessed.

7. Determine what to measure and how to measure it

How do you compare a koala to a tree fern? If all species were treated equally, then it would simply be the total number of each species, which would be a simple matter of addition (but only after the not insignificant task of estimating the number of each plants and animal species in Australia!). This is impractical for all but a few species (with the possible exception of birds). An alternative is the measurement of distinct habitats or ecosystems. This seems more practical and the extent of different vegetation types can at least be measured remotely for a plausible cost. The next question would then be: How to move from a simple measure of extent to one of quality? Reference condition metrics have been proposed for native vegetation (Gibbons *et al.* 2008; Cosier and McDonald 2010) but are still in the early stages of development.

8. Decide how much is enough for effective monitoring

You will never have enough resources to monitor every species or every metre of land, every year, to a level of ± 5 per cent, let alone ±1 per cent, of the true value. Identifying the critical areas for detailed investigation is the key. For biodiversity, remote sensing of land cover provides a relatively cheap way to cover large areas and for identifying places of greatest change to support targeting of on-ground monitoring efforts. Nested samples provide a way of mixing on-ground measurements with remote surveys and data on economic activity (from direct surveys or administrative data).

National monitoring will require national methods. It will be impossibly expensive to replicate the level of work done at some sites. For example, the amount of work done by David Lindenmayer and his colleagues in the Central Highlands of Victoria could not be replicated across the country. It could, however, be used to regularly benchmark remotely sensed data.

9. Be able to access and interpret data

There is little point collecting data if they cannot be accessed. Understanding who needs access to data and the required level of access (e.g. access to unit record files at one end and predefined summary tables at the other) is important. Many of us are familiar with the problem – we know the data exist but that these are not available to us for a variety of reasons. This is not just researchers seeking access to the information collected by public agencies, such as the ABS, but public agencies wanting access to the data that underpins the analyses of researchers.

One reason for the lack of data access is that data are usually hard-won and many people are reluctant to openly share data and instead see an opportunity to recoup some of the costs of data collection. A related problem, and one often experienced at the ABS, is that data are used without appropriate acknowledgement.

Some data are accessible, but not easily interpreted. Some data do not have an unambiguous 'good' direction. For example, in Australia, households' concern about environment issues fell from 75 per cent in 1992 to 57 per cent in 2004 (ABS 2004) (see Figure 9.2). Is this decline

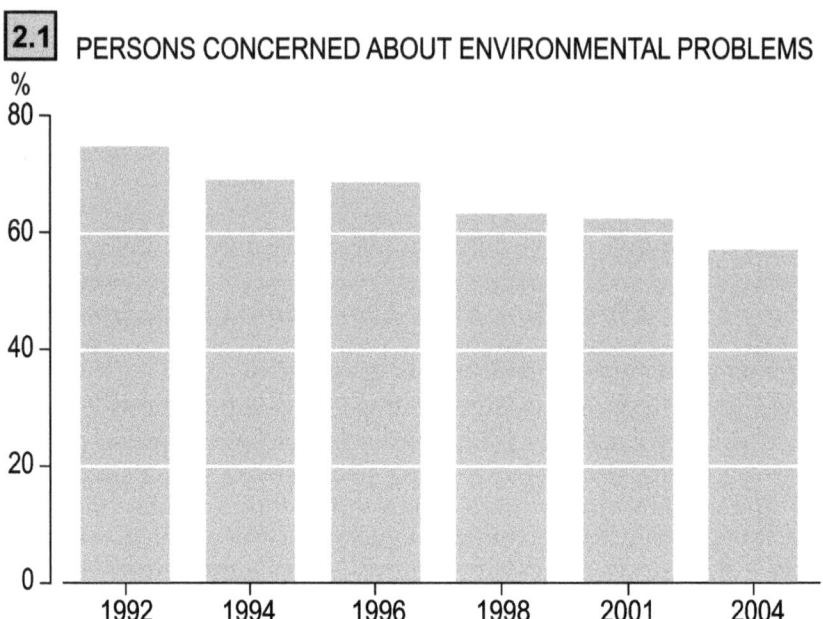

Figure 9.2. Persons concerned about environmental problems. (Source: ABS 2004).

good or bad? It could be good; people are not concerned because they think either the environment is improving or that the issues are being better addressed by governments, businesses and others in the community. The interpretation of the data collected and understanding what decisions the data may be used to inform, must be considered from the outset.

The provision of metadata (or the data about the data) is important for understanding the scope and quality of the data.

10. Define the right questions and maintain flexibility

There is little point having data on poorly conceived questions. Data collection programs need to have clear aims and questions to answer. The temptation is to ask narrowly defined questions of contemporary interest, the answers to which are mostly within your control. However, it is important to recognise that data may be able to answer a broader set of questions which are not at the forefront of present thought, but which may have been in the past or could be in the future. This can often be done with little or no extra cost. For example, landcover data collected as part of the national greenhouse gas inventory can be used for other purposes (e.g. land accounts).

It is therefore important to allow a degree of flexibility in monitoring programs, the addition of extra items and/or the deletion of redundant items. Allowing for the fact that data may be of use for purposes or users not envisaged at the outset will broaden the utility of these data. Recent natural disasters – floods and fire – have provided an opportunity to re-examine historical data and look to see how they might be used to increase our understanding of the natural world (see Gibbons, Chapter 19, this volume).

Conclusion

We are very good at identifying and analysing the problems related to environmental monitoring, but we've had few successes in systematically addressing them at broad scales, nationally and internationally. Much of the effort has been left to individual researchers or NGOs, with institutional structures and government agencies poorly equipped to put in place long-term monitoring programs.

In general, the focus has been on the scientific measurement of biodiversity and extremely little attention has been devoted to other aspects of biodiversity conservation, and in particular the cost-effectiveness of particular management interventions. The recognition of the need to link environmental and economic data, advances in the theory and practice of collecting environmental data, as well as recent academic and government initiatives give cause for cautious optimism.

Acknowledgements

I would like to thank the many people who have helped me to understand environmental accounting and the complexities of data collection and analyses and in particular: Chris Tidemann, Alessandra Alfieri, Rob Smith, Jeremy Webb, Ricardo Martinez-Lagunes, Jean-Louis Weber, Bram Edens, Michael Nagy, Neil Byron, Grahame Webb, John Ovington, Peter Comisari, Adam Sincock, Bob Harrison, Valdis Juskevics, and Andrew Cadogan-Cowper. Peter Comisari and Brendan Freeman provided valuable comments on this paper. Finally, thanks to David Lindenmayer and Phil Gibbons for their longstanding interest in environmental accounts and their relationship to basic data collection as well as the opportunity to be a part of this workshop.

Biography

Michael Vardon is currently Director of the Centre of Environment and Energy Statistics at the Australian Bureau of Statistics. He has been involved in the collection and analysis of environmental data for 20 years. This began with research and monitoring wildlife in south-eastern and northern Australia before a move to ABS in 2000 when he began working on environmental accounts. Between 2007 and 2009 he was an adviser to the United Nations Statistics Division on environmental accounts. He is currently a member of the SEEA Editorial Board and the Expert Group for the Framework for the Development of Environmental Statistics.

References

ABS (2000). 'Water Account, Australia 1993–94 to 1996–97'. ABS Cat. No. 4610.0. <http://www.abs.gov.au/ausstats/abs@.nsf/mf/4610.0> [Accessed 18/02/2011].

ABS (2004). 'Environmental issues: people's views and practices March 2004'. ABS Cat. No. 4602.0. <http://www.abs.gov.au/ausstats/abs@.nsf/mf/4602.0> [Accessed 18/02/2011].

ABS (2010*a*). 'Water Account, Australia 2008–09'. ABS Cat. No. 4610.0. <http://www.abs.gov.au/ausstats/abs@.nsf/mf/4610.0> [Accessed 18/02/2011].

ABS (2010*b*). 'Towards an integrated environmental-economic account for Australia, 2010'. ABS Cat. No. 4655.0.55.001. <http://www.abs.gov.au/ausstats/abs@.nsf/mf/4655.0.55.001> [Accessed 18/02/2011].

ANAO (2007). 'The conservation and protection of national threatened species and ecological communities'. Report No. 31, 2006–07, Performance Audit. <http://www.anao.gov.au/download.cfm?item_id=8B64F1991560A6E8AAB74 A9978383784&binary_id=9B0B8E391 560A6E8AAD7E4533DAEF3E9> [Accessed 10/02/2011] Australian National Audit Office, Canberra.

ANAO (2008). 'Regional delivery model for the Natural Heritage Trust and the National Action Plan for Salinity and Water Quality'. Report No. 21, 2007–08, Performance Audit.

<http://www.anao.gov.au/download.cfm?item_id=CE21 C4471560A6E8AA2773FE12C207
85&binary_id=F20BFC111560A6E8AA75D180CD56E65 2> [Accessed 10/02/2011]
Australian National Audit Office, Canberra.

Bureau of Meteorology (BoM) (2009). 'Pilot National Water Account'. <http://www.bom.gov.
au/water/nwa/nwaPilot.shtml> [Accessed 18/02/2011].

Cosier P and McDonald J (2010). 'A common currency for building environmental
(ecosystem) accounts'. Paper presented at the 16th London Group meeting. Santiago,
25–28 October 2010. <http://unstats.un.org/unsd/envaccounting/londongroup/meeting16/
LG16_22a.pdf> [Accessed 18/02/2011].

Gibbons P, Briggs SV, Ayers DA, Doyle S, Seddon J, McElhinny C, Jones N, Sims R and Doody
JS (2008). Rapidly quantifying reference conditions in modified landscapes. *Biological
Conservation* **14**, 2483–2493.

Royal Swedish Academy of Science (1984). The Prize in Economics 1984 for Pioneering Work
in the Development of Systems of National Accounts – Press Release. <http://nobelprize.
org/nobel_prizes/economics/laureates/1984/press.html>

Smith R (2007). The development of the SEEA 2003 and its implementation. *Ecological
Economics* **61**, 592–599.

Statistics Canada (2002). 'Statistics Canada's quality assurance framework'. <http://www.
statcan.gc.ca/bsolc/olc-cel/olc-cel?lang=eng&catno=12-586-X> [Accessed 18/02/2011].

United Nations (1993). 'System of integrated environmental and economic accounting'.
United Nations, New York.

United Nations (2003). 'System of integrated environmental and economic accounting'.
United Nations, New York. <http://unstats.un.org/unsd/envaccounting/seea2003.pdf>
[Accessed 18/02/2011].

United Nations (2007). 'System of environmental economic accounting for water'. United
Nations, New York. <http://unstats.un.org/unsd/statcom/doc07/SEEAW_SC2007.pdf>
[Accessed 18/02/2011].

United Nations (2009). 'Report on the global assessment of water statistics and water
accounting'. UN Statistical Commission Paper. <http://unstats.un.org/unsd/statcom/
doc09/BG-WaterAccounts.pdf> [Accessed 18/02/2011].

United Nations (2010). 'International recommendations for water statistics'. United Nations,
New York. <http://unstats.un.org/unsd/statcom/doc10/BG-WaterStats.pdf> [Accessed
18/02/2011].

Vardon M, Lenzen M, Peevor S and Cresser M (2007). Water accounting in Australia.
Ecological Economics **61**, 650–659.

Vardon M, Martinez-Lagunes R, Nagy M and Gan H (2012). The system of environmental
economic accounting for water: development, implementation and use. In *Water
Accounting: International Approaches to Policy and Decision-making.* (Eds J Godfrey and
K Chalmers) pp. 32–57. Edward Elgar, UK.

Victorian Government (2011). BushTender. <http://www.land.vic.gov.au/DSE/nrence.nsf/Link
View/15F9D8C40FE51BE64A256A72007E12DC37EBE3A50C29F4F8CA2573B6001A8
4D5> [Accessed 15/02/2011].

Weber J-L (2007). Implementation of land and ecosystem accounts at the European
Environment Agency. *Ecological Economics* **61**, 695–707.

GOVERNMENT AGENCY AND NGO PERSPECTIVES

10 COWS, COCKIES AND ATLASES: USE AND ABUSE OF BIODIVERSITY MONITORING IN ENVIRONMENTAL DECISION-MAKING

James Fitzsimons

Lesson #1. Good long-term monitoring makes for informed and confident decisions on land management.

Lesson #2. Monitoring showing species and habitat decline can directly lead to better protection mechanisms.

Lesson #3. Results of monitoring can be ignored, misused and misquoted to achieve political ends.

Lesson #4. Are we seeing a decline in systematic species surveys by government?

Lesson #5. We don't know enough about what monitoring is happening and why monitoring isn't happening.

Lesson #6. Disparate data sets and cumbersome collection methods are hindering species status monitoring.

Lesson #7. Make better use of existing resources and expertise.

Lesson #8. Make monitoring data more accessible and enable it to be more repeatable.

Lesson #9. Embed the requirement for monitoring in biodiversity and threatened species legislation.

Lesson #10. Understand better the social elements of ecological monitoring.

Introduction

Ecological monitoring and the collection of ecological data are essential components of environmental management. Governments, NGOs and philanthropists are all increasingly insisting on monitoring programs that demonstrate the outcomes of funded conservation activities. Monitoring of biodiversity conservation outcomes can be carried out in various forms including status measures to strategy effectiveness measures.

This chapter focuses largely on the conditions that enable and encourage (or conversely, hinder) monitoring. It draws on examples and observations from my work in conservation and protected area policy and planning, in government departments and agencies and in the non-government sector. Defining 'successes in monitoring' could be interpreted in a number of ways, including (1) being able to maintain a monitoring program in the long-term; (2) being able to demonstrate a conservation success (e.g. documenting the recovery of a species); or (3) monitoring leading to a positive change in management practice, law or policy etc. In this chapter I touch mostly on (1) and (3) and outline two examples of monitoring where *both* success and failure is evident (i.e. alpine grazing and electronic species databases).

Lessons

1. Good long-term monitoring makes for informed and confident decisions on land management (Alpine Grazing part 1)

Much progress has been made in the use and consideration of the results of ecological monitoring in informing land use decision-making. Here, I focus on the case of cattle grazing in the Australian Alps and specifically Victoria's Alpine National Park. Cattle grazing in the Alps has been a controversial issue for many years, incorporating debates about cultural heritage, nature conservation and national park objectives. Monitoring the impact of grazing on the high country began in 1945, when botanist Maisie Fawcett fenced out some plots on the Bogong High Plains to look at the impact of cattle grazing on the plant communities. These have been monitored continuously since that time. Despite the creation of the Alpine National Park in 1989, cattle grazing remained, even though the practice was at odds with the objectives of the park. In 2005, the government-appointed Alpine Grazing Taskforce drew heavily on the long-term monitoring initiated by Fawcett (e.g. Wahren *et al.* 1994; Williams *et al.* 1997) in their finding that grazing was not optimal for nature conservation, nor soil and water quality, and that there was no evidence for a reduction in the likelihood of fire due to grazing (Alpine Grazing Taskforce 2005). The Maisie Fawcett plots were visited by the Taskforce and were a powerful visual demonstration of the differences between grazed and ungrazed sites. The government subsequently removed cattle from the park in 2005, again citing the compelling results of long-term monitoring: 'There is overwhelming scientific evidence that cows as hard-hooved heavy animals are damaging the sensitive Alpine environment' (Bracks and Thwaites 2005). (However, see Lesson #3 below).

2. Monitoring showing species and habitat decline can directly lead to better protection mechanisms

The south-eastern red-tailed black-cockatoo (*Calyptorhynchus banksii graptogyne*) is an endangered subspecies of parrot occurring in south-west Victoria and south-east South Australia where it relies on brown (*Eucalyptus baxteri*) and desert stringybarks (*E. arenacea*) and bulokes (*Allocasuarina luehmannii*) for feeding, as well as eucalypts such as river red gums (*E. camaldulensis*) for nesting. Bulokes occur mainly as paddock trees in the largely cleared west Wimmera region of Victoria. In 2006, a survey of buloke trees remaining in west Wimmera landscapes was undertaken (Maron and Fitzsimons 2007). This was a repeat of a survey undertaken originally by Maron (2005) on rates of loss between 1981–82 and 1997. Maron (2005) recorded substantially lower annual rates of tree loss for the 15-year period to 1997 than we found in the eight years to 2005, despite native vegetation clearing controls being in place since 1989, and more recent policy intended to provide some protection for paddock trees having been introduced in 2002. This loss was mainly due to increase in centre pivot irrigation (Maron and Fitzsimons 2007).

The publication of these data, along with community pressure which led to Ministerial pressure on the local council, the red-tailed black-cockatoo recovery team, and a desire to avoid all clearing applications ending up in a planning disputes tribunal, resulted in planning scheme amendments and overlays which specifically recognised live buloke trees (M. Maron pers. comm. 2011). The effectiveness of the approach remains to be seen (Maron *et al.* 2010). Although monitoring in this instance is considered a success as it strengthened protective mechanisms, continual monitoring of both the cockatoos and their habitat will be required, especially considering past changes to policy/protective mechanisms did not halt habitat decline.

3. Results of monitoring can be ignored, misused and misquoted to achieve political ends (Alpine Grazing part 2)

Despite the removal of cattle from the Alpine National Park in 2005 (see Lesson #1 above), the Victorian opposition, in the lead-up to the 2010 Victorian election, included in their policy a plan to reintroduce cattle to the high country (Walsh 2010). While conservationists and ecologists were clearly disappointed that reintroduction of a known threatening process to a protected area had been proposed, they felt the line 'Cattle grazing can be an important tool to reduce fire risk *where appropriate* on Crown land' [emphasis added] would rule this action out considering the extent of and well-documented past monitoring on the issue. However, after winning the election, it was announced that on 12 January 2011 the 'Coalition Government is trialing strategic cattle grazing as a tool to mitigate bushfire risk in Victoria's high country, delivering on its election commitment' (Smith 2011). Cattle were allowed back into the park on the same day. Despite the findings of the Alpine Grazing Taskforce (2005) and subsequent research (e.g. Tolsma 2006; Williams *et al.* 2006) that, based on significant monitoring, grazing was both detrimental to alpine biodiversity and does not 'reduce blazing', the government cited the need for more science:

- 'This scientific trial … will provide much-needed evidence on the effectiveness of strategic cattle grazing for fuel and fire management purposes.'
- 'The Coalition Government is committed to making transparent and informed decisions on bushfire management in Victoria's high country based on credible scientific evidence.'
- 'Current information on the effect of cattle grazing for bushfire mitigation is limited' (Smith 2011).

However, it was subsequently reported that the independent scientist commissioned to undertake the research was unaware that cattle had been released into the park on that date, that no benchmark data had been collected and that any 'monitoring' would not begin until late 2011 (Morton 2011). Nor had research approvals been sought through the usual channels of the park management agency. Disregarding advice based on long-term monitoring and clear trends is not uncommon in government. However, the badging of blatantly political decisions as a 'need for more science' threatens to compromise biodiversity monitoring in a similar fashion to the 'debate' on climate science.

4. Are we seeing a decline in systematic species surveys by government?

Although undertaking inventories or including non-systematic species records into state-based databases is not monitoring *per se*, collectively these data can be used to monitor the range or population status of a species over time. For example, atlas data has been a key data set in monitoring the status of birds across the country as evidenced by its use in action plans and in threatened species nominations to governments. The 2008 Victorian State of the Environment report (CES 2008) highlighted that, while fluctuations in entries for flora and fauna occurred over time in that state (with peaks for government scientists undertaking systematic surveys of particular areas), there has been a significant decline in the number of records entered into the respective databases in recent years (see Figure 10.1a–c). Conversely, while there are peaks in data collection during periods where the two Atlas of Australian Birds reports were being compiled, there has been a consistent survey effort since the early 2000s, highlighting the potential importance of citizen science in collecting essential data at a large scale (the records were collected largely by volunteers) (see Figure 10.1a–c).

The broad (and at times unclear) nature of the trends reported for threatened species that are being actively monitored in Victoria (see Tables LB3.5 and LB3.6 in CES 2008) suggests that either monitoring techniques and effort are not adequate to determine trends for many species, or that longer time-frames for monitoring are required. Although a lack of adequate resources is likely to be a key factor, other elements may also influence the implementation of successful techniques and timeframes (see Lessons #5 and #10 below).

5. We don't know enough about what monitoring is happening and why monitoring isn't happening

Private land conservation mechanisms have the potential to make a significant contribution to biodiversity conservation in Australia (Fitzsimons and Wescott 2001, 2008). In a review of conservation outcomes of conservation covenanting programs across Australia, Fitzsimons and Carr (2007) found that the role of monitoring and types of monitoring varied widely. For example, monitoring programs ranged from the basic statewide to regional inventories, such as number and area of covenants and increase in growth in signing covenants per year, through to assessments of the contribution that covenants are making to the conservation estate at the bioregional level (e.g. enhancing representation and/or improving linkages in the landscape or buffering protected areas). Other monitoring measures included site-based assessments such as complying with the conditions of the covenant and various forms of ecological monitoring. Some programs did all of these, whereas others only undertook the broader assessment.

In terms of on-ground ecological monitoring, the techniques and emphasis between programs varied and the purpose for doing this was more to inform management than to necessarily gain quantifiable ecological data suitable for statistical analysis. Some were using methods that were consistent or comparable with what was being used in the rest of the jurisdiction (i.e. elsewhere with the state nature conservation agency/parks service), unlike others that had a more simplified or more advanced version of what is used elsewhere in the state. Some covenant programs had collected benchmark ecological information for most covenants at the time of signing and most programs now undertake this on the signing of new covenants. Site visits ranged from yearly to five-yearly or on an 'as-needs' basis. A lack of resources to monitor (staff numbers and time), knowing what to monitor, inconsistent monitoring methodologies, lack of benchmark data and length of time to see meaningful results from monitoring, were all considered potential barriers to evaluating the biodiversity conservation outcomes of conservation covenants.

6. Disparate data sets and cumbersome collection methods are hindering species status monitoring

There are significant volumes of ecological data sitting on the computers, in filing cabinets and notebooks of ecologists, government agencies and the general public. Part of the reason that species record data, for example, is not more readily shared is the somewhat cumbersome process of having to fill out hardcopy data record sheets that still persist for many atlas schemes. Birds Australia's Atlas has made significant advances in making it easy for most people to enter monitoring surveys remotely and electronically via Birdata (www.birdata.com.au). However, the rapid evolution of technologies such as smart phones, GoogleEarth, digital cameras with

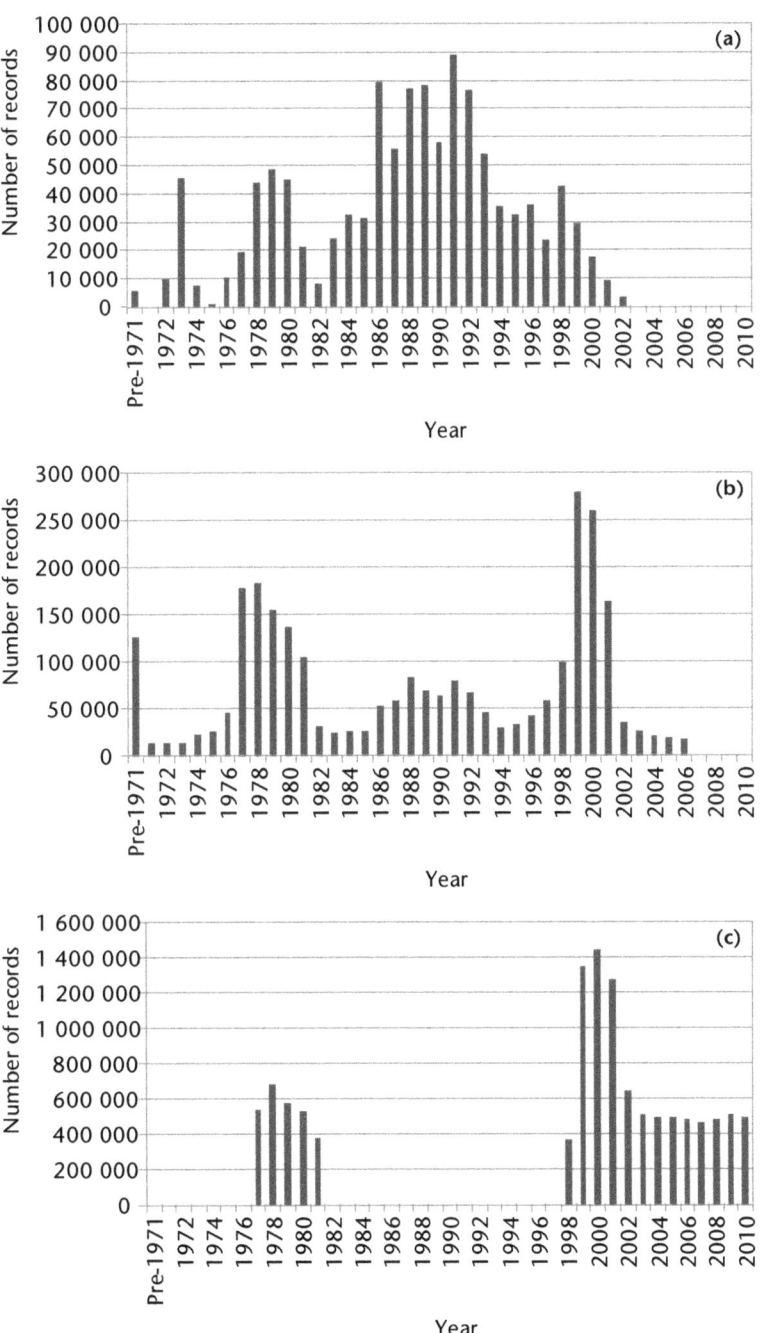

Figure 10.1. Numbers of records entered each year for (a) flora and (b) fauna in databases maintained by the Victorian Government and (c) in Birds Australia's *Atlas of Australian Birds*. Date ranges: Victorian flora records up to 2003, Victorian fauna records up to 2006, bird atlas data 1977–2010. Note fluctuations and recent decline for entries in Victoria and consistency of bird records in the 2000s. Small numbers of records did occur between 1982 and 1997 for the *Atlas of Australian Birds* (A. Silcocks pers. comm. 2011). Figure adapted from Gullan (2003), CES (2008) and 2011 data from Birds Australia.

GPS etc., which have the potential for making the recording, capture and dissemination of ecological data easier for a far greater number of people, mean that atlas repositories will need to be tailored and flexible.

The current disparate and sometimes duplicated species atlases and databases of state conservation agencies, museums and other government and non-government scientific institutions urgently require incorporation into a seamless and continually updated national atlas. The Atlas of Living Australia (www.ala.org.au) promises to go part of the way to achieving this.

7. Make better use of existing resources and expertise

There will never be enough resources or people to monitor everything that should be monitored. We need to better harness and direct people that do have the time, expertise, and/or self or subsidised finances to do it. For example, links between universities and land management agencies, while improving, really have not taken full advantage of keen and enquiring minds of undergraduate and postgraduate students. Likewise, while atlas projects have secured a huge amount of valuable data to enable monitoring, further guidance and encouragement from atlas coordinators could improve this process. For example, explicitly highlighting and encouraging surveys of under-represented sites and species would not only improve the comprehensiveness of these approaches to atlassing but potentially provide a greater sense of contribution from participants.

8. Make monitoring data more accessible and enable it to be more repeatable

Do we provide enough information in our communication of monitoring to allow surveys to be replicated by others? In academic journal articles (the most commonly utilised public output for the results of ecological monitoring), study sites are often described in broad terms and results consolidated to show trends. While this is largely driven by the journals, the inclusion of detailed information on survey sites, methods and results in online 'supplementary material' or in publicly accessible reports or websites should be a requirement for all surveys or monitoring programs funded by the public purse.

While publishing in peer-reviewed journals is essential for scientific credibility, it is slow, rarely *freely* accessible to the majority of the population, and generally rarely read by politicians, bureaucrats and natural resource managers. Providing freely available, easy to read summaries of the outcomes of monitoring via the web or 'glossy booklets' (e.g. Fitzsimons *et al.* 2010) is essential to reach a wider audience and increase the impact of biodiversity monitoring.

Unfortunately, in some cases, the release of monitoring data collected by government scientists or as part of consultancies to government is hindered and sometimes may never see the light of day, often when it presents 'bad news' on a species' status or failure of policy (e.g. ongoing habitat loss).

9. Embed the requirement for monitoring in biodiversity and threatened species legislation

Requirements for monitoring in policy and legislation are still weak, even for threatened species where recovery efforts rely on monitoring to determine species trends. Incorporating the need for monitoring into legislation is not foolproof – there are many examples where legislative requirements are not fulfilled by government, often with little or no legal or political ramifications. Nonetheless, fulfilling legislative requirements are usually priorities for bureaucracies, and remain core business when finances are tight.

10. Understand better the social elements of ecological monitoring

Why do some birdwatchers get involved in atlassing and others don't? If citizen science is to be used to its greatest potential, a better understanding of the interests, attitudes, abilities and restrictions of those involved and those *not* involved is needed. Work by Weston *et al.* (2006) and Wolcott *et al.* (2008) has started to document some of this, but clearly more work is needed.

But what about ecologists and environmental managers in academic institutions, in government agencies and in NGOs? While many (or most?) of these recognise monitoring is a good thing, if not essential, it is often not given the priority it deserves. How much of this is due to the *personal* interest or value placed on monitoring by various lines of management, as distinct from policy or resourcing? These are important areas of social research that should be addressed.

Conclusions

Long-term ecological monitoring can provide clear and compelling information for decision-makers. Often decision-makers can choose to accept this and use the credibility of science to sell the case for controversial decisions. However, this information can also be ignored, or claimed to be not 'complete enough'. The issue of grazing in Victoria's Alpine National Park presents a case in point. Ensuring results of monitoring are freely available and easily accessible in the public domain makes claims of 'not enough data' harder to defend. At the same time, scientists should not be afraid to speak out where monitoring data have been misquoted, misused or ignored by governments to justify political decisions.

There will never be enough resources to monitor all the things that need to be monitored and we need to harness more creative solutions. But we also need a better understanding of the social and cultural barriers to monitoring. There are many more ways in which monitoring and the conditions that encourage monitoring could be improved as outlined in the other chapters of this book.

Acknowledgements

Thanks to Martine Maron for information on responses to buloke clearing in the Wimmera, Andrew Silcocks for supplying data from the Atlas of Australian Birds, the (then) Department of the Environment and Water Resources for funding the covenanting research, and Janelle Thomas, David Lindenmayer and Phil Gibbons for comments on a draft of the chapter.

Biography

James Fitzsimons is the Director of Conservation for The Nature Conservancy's Australia Program and an honorary research fellow at the School of Life and Environmental Sciences, Deakin University. His particular research interests are in the fields of protected area policy, practical conservation planning and mechanisms to integrate conservation outcomes on public and private lands. He has worked in the fields of conservation policy and planning for government environment departments and agencies, and for non-government environment organisations.

References

Alpine Grazing Taskforce (2005). 'Report of the investigation into the future of cattle grazing in the Alpine National Park'. Department of Sustainability and Environment, Melbourne.

Bracks S and Thwaites J (2005). 'High country grazing continues outside national park'. Media Release from the Premier and the Minister for Environment, 24 May 2005, Available: <http://www.legislation.vic.gov.au/domino/Web_Notes/newmedia.nsf/bc348d5 912436a9cca256cfc0082d800/4fbb24d02982648eca25700c0019adfd!OpenDocument> [Accessed 19 October 2011].

Commissioner for Environmental Sustainability (CES) (2008). 'State of the Environment Victoria 2008'. Commissioner for Environmental Sustainability, Melbourne.

Fitzsimons J and Carr B (2007). 'Evaluation of the effectiveness of conservation covenanting programs in delivering biodiversity conservation outcomes'. Report for the Australian Government Department of Environment and Water Resources. Bush Heritage Australia, Melbourne.

Fitzsimons J, Legge S, Traill B and Woinarski JCZ (2010). 'Into Oblivion? The disappearing native mammals of northern Australia'. The Nature Conservancy, Melbourne. Available: <http://www.nature.org/ourinitiatives/regions/australia/explore/ausmammals.pdf> [Accessed 19 October 2011].

Fitzsimons J and Wescott G (2001). The role and contribution of private land in Victoria to biodiversity conservation and the protected area system. *Australian Journal of Environmental Management* **8**, 142–157.

Fitzsimons JA and Wescott G (2008). Ecosystem conservation in multi-tenure reserve networks: The contribution of land outside of publicly protected areas. *Pacific Conservation Biology* **14**, 250–262.

Gullan P (2003). The Victorian Flora Information System 1975–2003. Unpublished report. Viridans Biological Databases, Melbourne.

Maron M (2005). Agricultural change and paddock tree loss: Implications for an endangered subspecies of Red-tailed Black-Cockatoo. *Ecological Management & Restoration* **6**, 206–211.

Maron M and Fitzsimons JA (2007). Agricultural intensification and loss of matrix habitat over 23 years in the West Wimmera, south-eastern Australia. *Biological Conservation* **135**, 603–609.

Maron M, Dunn PK, McAlpine CA and Apan A (2010). Can offsets really compensate for habitat removal? The case of the endangered red-tailed black-cockatoo. *Journal of Applied Ecology* **47**, 348–355.

Morton A (2011). 'Grazing study can't start for months'. *The Age*, 29 January 2011. Available: <http://www.theage.com.au/victoria/grazing-study-cant-start-for-months-20110128-1a8fx.html> [Accessed 23 March 2011].

Smith R (2011). 'Research begins on strategic cattle grazing to reduce bushfire risk'. Media release from the Hon Ryan Smith MP, Minister for Environment and Climate Change. 12 January 2011. Available: <http://premier.vic.gov.au/2011/01/research-begins-on-strategic-cattle-grazing-to-reduce-bushfire-risk/> [Accessed 23 March 2011].

Tolsma A (2006). 'Alpine (livestock) grazing position statement by the Ecological Society of Australia'. Available: <http://www.ecolsoc.org.au/documents/ESAFinalPositionStatementonAlpineGrazingSept20061.pdf> [Accessed 23 March 2011].

Wahren CH, Papst WA and Williams RJ (1994). Long-term vegetation change in relation to cattle grazing in subalpine grassland and heathland on the Bogong High Plains: an analysis of vegetation records from 1945 to 1994. *Australian Journal of Botany* **42**, 607–639.

Walsh P (2010). 'Vic Coalition to reinstate high country cattle grazing to reduce fire risk'. Victorian Liberal Nationals Coalition Media Release. 9 January 2010. Available: <http://vic.liberal.org.au/LinkClick.aspx?fileticket=/bElf7kQ56M=&tabid=189> [Accessed 23 March 2011].

Weston MA, Silcocks A, Tzaros C and Ingwersen D (2006). A survey of contributions to an Australian bird atlassing project: Demography, skills and motivation. *Australian Journal on Volunteering* **11**, 51–58.

Williams RJ, Papst WA and Wahren CH (1997). 'The impact of cattle grazing on the alpine and sub-alpine plant communities of the Bogong High Plains'. Report to the Victorian Department of Natural Resources and Environment. Department of Natural Resources and Environment, Melbourne.

Williams RJ, Wahren C-H, Bradstock RA and Müller WJ (2006). Does alpine grazing reduce blazing? A landscape test of a widely-held hypothesis. *Austral Ecology* **31**, 925–936.

Wolcott I, Ingwersen D, Weston MA and Tzaros C (2008). Sustainability of a long-term volunteer-based bird monitoring program: Recruitment, retention and attrition. *Australian Journal on Volunteering* **13**, 48–53.

11 A PARK MANAGER'S PERSPECTIVE ON ECOLOGICAL MONITORING

Tony Varcoe

Lesson #1. Over the past decade there has been a global revolution in Protected Area monitoring and management effectiveness frameworks.

Lesson #2. An increasing number of conservation managers have been developing and testing new monitoring programs that are very management focused.

Lesson #3. Insufficient effort has been devoted to planning and design of monitoring programs to maximise benefits and minimise risk.

Lesson #4. Monitoring activities have rarely been directly connected to clear management objectives and questions.

Lesson #5. There have been limited examples where monitoring results have been well communicated to the different required audiences and fed into the adaptive management cycle.

Lesson #6. Build and support an evaluation culture in park and land managers.

Lesson #7. Develop a monitoring framework with a hierarchy of monitoring to reflect different management aims.

Lesson #8. Develop a monitoring plan.

Lesson #9. Use available tools such as conceptual models and risk assessments to make more informed decisions about what to monitor.

Lesson #10. Build truly collaborative monitoring partnerships.

Introduction

How 'healthy' are our parks? How do we know if, and to what extent, our ecosystem and park management objectives are being met? How do we know whether our management actions are effective? These obvious and legitimate questions are fundamental to the business of park management. Yet they have often been difficult to answer despite a long history of data collection by scientists, land managers, naturalists and the community. Increasingly, governments and the community are expecting park managers to report on the effectiveness of Protected Areas in conserving biodiversity and other park values (e.g. Parrish *et al.* 2003; Hockings *et al.* 2005). Conversely, a lack of evidence-based reporting on effectiveness has the potential to compromise outcomes and jeopardise the investment made in Protected Areas for conservation (Cook *et al.* 2009). Importantly, while adaptive management is espoused as the standard approach for today's park managers, building and applying evidence to inform park management decisions requires carefully targeted monitoring, based on well-defined objectives that are directly relevant to the needs of park managers.

In company with numerous park and land management agencies around the world, the experience of Parks Victoria has shown that, despite undertaking a considerable amount of activity in ecological monitoring over the past decades, there are limited examples where long-term monitoring programs can demonstrate changes both in resource condition and the effectiveness of management actions in achieving desired objectives (e.g. Parks Victoria 2007; VAGO 2010). Notable exceptions include monitoring of introduced and native herbivores and their impacts in Victoria's mallee parks (Cheal *et al.* 2007), vegetation monitoring in alpine ecosystems (e.g. Wahren *et al.* 2001) and monitoring of fauna in the wet forest ecosystems of the Central Highlands (Lindenmayer 2009). Some of the most commonly reported barriers to effective monitoring include: lack of clear ecological objectives and targets to measure against; lack of, or vague, monitoring questions; insufficient planning and design; inconsistent methods and data standards; projects that fade away with changing resourcing priorities; park managers and scientists not working in tandem; and poor information systems (see Legg and Nagy 2006; Lindenmayer 2009).

Much has been learnt over the past decade and there is now a growing recognition that ecological monitoring efforts need to be more focused to reflect land manager needs. Land managers also need to understand that good science takes commitment.

Successes

1. Over the past decade there has been a global revolution in Protected Area monitoring and management effectiveness frameworks

Following the hard lessons learnt from monitoring program failures, over the past decade there has been a revolution in developing more strategic and robust frameworks for management-focused ecological monitoring in Protected Areas. Initiatives such as Parks Canada's Ecological Integrity Monitoring (see Woodley 2010), the United States National Parks Service (US NPS) Vital Signs monitoring (see Fancy *et al.* 2009), the New Zealand National Heritage Management System (see Lee *et al.* 2005), the South African Kruger Model of adaptive management (Biggs and Rogers 2003), as well as locally developed programs such as the Living Murray monitoring program (see MDBC 2008) have all been developed using strong scientific principles with clearer objectives and resourcing commitments.

Likewise there has been enormous growth in planning and implementation of management effectiveness evaluation processes, most notably through the World Commission on Protected Areas (WCPA) Science and Management theme (see Leverington and Hockings 2004). Many countries in almost all continents are now either developing or implementing more systematic evaluation of management effectiveness. The WCPA model for evaluating management effectiveness of parks has been adopted by both Parks Victoria and the Office of Environment and Heritage through their State of the Parks programs (see Parks Victoria 2007).

2. An increasing number of conservation managers have been developing and testing new monitoring programs that are very management focused

Building on the lessons learnt from international and national monitoring programs, a number of Australian public and private conservation management organisations are currently rolling out new management-focused ecological monitoring programs. Parks Victoria is piloting its new objective-based monitoring program. While still in its infancy, 'Signs of Healthy Parks' is seeking to implement a more robust, strategic and targeted approach to ecological monitoring

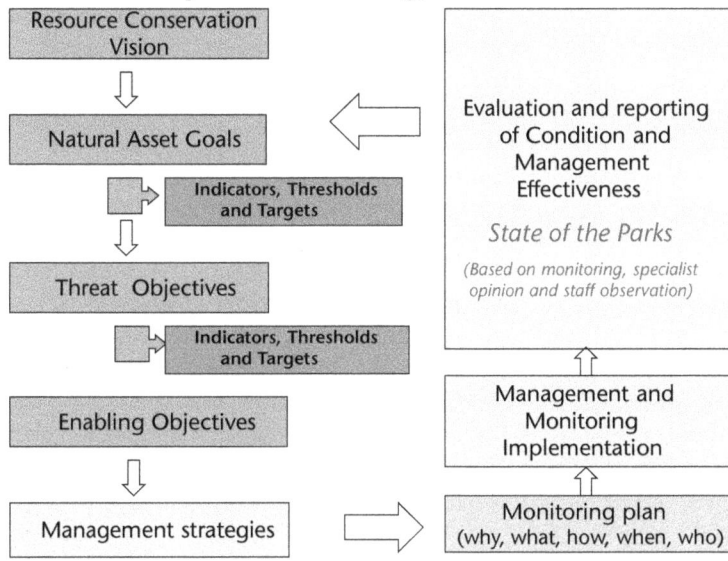

Figure 11.1. Parks Victoria's objectives hierarchy for adaptive management.

than in the past. While Parks Victoria recognises the need for a level of broader-based monitoring surveys to improve knowledge about the presence or absence of park values, 'Signs of Healthy Parks' will focus largely on informing management questions and evaluating the extent to which park objectives are being met. Additionally, we have initiated a number of Adaptive Experimental Management (AEM) programs to test the effects of specific management actions including ecological thinning, fox control and weed control. Notably there is also increased recognition of the need for question-based monitoring in newly established fire management monitoring programs being developed by sister agencies such as the Department of Sustainability and Environment (DSE).

In recent years, there has been growing recognition that clear management objectives are the foundation of effective management and monitoring programs. Parks Victoria has recently commenced a fast-track process to develop more measurable conservation objectives for its major parks, including development of an 'objectives hierarchy' (Figure 11.1). This is adapted from the South African Kruger National Park objectives hierarchy and will be the backbone of all new park management plans. There is still much work to be done in this area. Nationally, a number of park management agencies (e.g. Tasmania, New South Wales) are developing new ecological monitoring and management effectiveness frameworks.

Failures

3. Insufficient effort has been devoted to planning and design of monitoring programs to maximise benefits and minimise risk

For any wise investor, thorough planning and 'doing the sums' is the key to maximising returns and minimising losses. Well-designed and targeted ecological monitoring should produce significant management dividends by providing the evidence base to adjust management actions and priorities. The risk of poorly designed monitoring is that it may not be able to answer the management question while wasting valuable resources.

With such a variety of demands in park management, Parks Victoria, like many other Australian park agencies, has a large proportion of staff who are 'generalists'. Parks Victoria's own experience in developing and implementing scientific monitoring protocols shows that while many field staff appreciate the need for ecological monitoring, they are often challenged by the need for strict scientific standards in design and implementation. Many operational staff and managers (as well as volunteer groups) have little experience in applying the experimental design concepts of certainty, replication and detectability and they therefore need a high level of scientific support to implement monitoring programs to the standards required by protocols. Indeed, there have been examples where robust scientific design has been perceived as an obstacle created by the scientists to impede the monitoring desires of park rangers.

It is a harsh lesson that failure to undertake the necessary assessment of the effort and costs required to detect a certain level of change is one of the major reasons for monitoring programs failing to provide meaningful results to park managers. This problem can reinforce a perception that monitoring isn't worth the effort. The two areas of greatest need in ecological monitoring skills are planning and design at the 'front end' and analysis and interpretation of monitoring data at the 'back end'. The issue of developing robust design still remains one of the biggest obstacles in establishing new programs and improving existing programs. There is, however, a growing recognition that the inclusion of appropriate analysis as part of the planning and design phases of monitoring programs needs to be an important consideration in all new programs.

To address the above issues, Parks Victoria has increased its efforts to train operational staff on the basic scientific principles of ecological monitoring (Parks Victoria 2009), and we have built much improved partnerships between park managers and scientists through the Research Partners Program. Nevertheless, a significant challenge exists to find the right balance between building internal science capacity to enable park managers to 'ask the right questions' and effectively enabling access to those essential scientific and specialist ecological skills from external sources.

4. Monitoring activities have rarely been directly connected to clear management objectives and questions

Clear and measurable ecological objectives should be the foundation for targeted monitoring. Without clear objectives, monitoring is unlikely to be able to inform the fundamental management question of 'How are we going?' Many authors have documented the importance of linking monitoring programs to specific management needs (e.g. Nichols and Williams 2006; Stevens *et al.* 2005; Field *et al.* 2007). Our experience is that of all of the components of monitoring, the greatest amount of discussion and effort is dedicated to eliciting clear monitoring questions based around meaningful management objectives. While park management plans have been designed to provide strategic directions across a wide range of issues, they have often provided goals and objectives that are somewhat nebulous. Measuring outcomes against these nebulous goals has been very difficult. It is indeed difficult to connect monitoring questions to measurable management objectives if those objectives don't exist. The recent experience of other Protected Area agencies such as Parks Canada's Ecological Integrity (EI) framework provides valuable insight into translating broad nature protection goals into more specific and measurable endpoints based on desirable ecological conditions (Woodley 2010).

At the more measurable end of objectives, the development of meaningful and credible targets and 'thresholds of concern' to compare monitoring results against is a substantial conservation management challenge. While thresholds and targets are well advanced in resource management areas such as water, many park and other land managers fall into decision paralysis when required to develop similar measures for biodiversity assets on the basis that there is insufficient evidence. It is essential that park managers work closely with ecological specialists in developing targets and thresholds, recognising that, in the face of imperfect data, we might need to take some risks and learn as we go.

5. There have been limited examples where monitoring results have been well communicated to the different required audiences and fed into the adaptive management cycle

If a primary aim of ecological monitoring is to inform management responses, one of the most important, but least recognised needs, is the ability of scientists to tell evidence-based stories. Providing often complex information to managers in clear and easily understood formats is fundamental to those managers 'owning' and applying the results from monitoring. Much work has been recently undertaken by Parks Canada and the US NPS to develop frameworks for matching stakeholder needs to different types of monitoring information.

Understanding who the audiences are, and what type of information they seek, must strongly influence the way that monitoring data are used. In developing 'Signs of Healthy Parks', we are primarily targeting park managers to provide them with an improved evidence-base to inform their management actions and management priorities. However, with care, the information collected can and should be used to provide reliable and meaningful information at the aggregated scale. Such information can inform 'State of the Parks' reporting to managers, policy makers and the community on the overall 'health' of our parks and effectiveness of park management at regional and statewide scales.

Ecological monitoring programs must be more than a time series of data. Just as lack of monitoring cannot close the adaptive management loop, inadequate interpretation and discussion of management implications do not complete the monitoring loop. While publication of monitoring data in peer-reviewed research journals is an important indicator of scientific performance, sitting down with park managers to explain the results and potential applications of these data also should also be essential.

Finally, while condition 'report cards' have become a standard reporting tool in many land management agencies, users of this information need to look beyond the traffic lights and arrows and ask 'What is the information based on?' Agencies such as Parks Canada and US NPS have set a new standard for this type of reporting, recognising the direct link between raw data and aggregated reporting (see Fancy *et al.* 2009; Woodley 2010).

Solutions

6. Build and support an evaluation culture in park and land managers

Organisational culture is a strong driver of whether monitoring programs are perceived as being either a fundamental part of management or a 'would like to do' activity. Evolving a culture from being largely activity focused (where success is measured by *more* activity) into a culture that recognises evaluation as an essential component of normal business, requires clear vision and direction, logical steps, user-friendly tools, access to the right skills and time.

For ecological monitoring programs to become more effective, organisations and individuals including park staff and their managers, need to believe that monitoring can actually make a difference to management. Indeed, well-designed and targeted monitoring programs that provide direct feedback on achievement of management objectives would seem a logical and essential component of all good project management. So why hasn't management-focused monitoring become part of normal business? One of the biggest barriers is the perception that monitoring is an end in itself rather than being a practical tool to provide evidence for adaptive management. With limited resources and many demands in park management, it is entirely legitimate that park managers should be satisfied that monitoring will actually assist them to make more informed decisions rather than be a never-ending resource hungry monster.

Gathering and applying information from well-targeted monitoring programs also requires patience and commitment. Experience in developing a number of adaptive experimental management programs in Victoria's parks (e.g. Robley *et al.* 2008) has shown that keeping up the commitment of resources for monitoring is a major challenge. There have been numerous examples where well-meaning people have initiated monitoring programs only to watch them fade away as the next priority takes over resources and energy. With so many demands to report achievements in annual and three-year cycles, building organisational culture that understands the longer time frames required for demonstrating monitoring results is a significant challenge.

It is equally important that park management organisations develop a culture that is prepared to deal with bad news in the form of monitoring results that show declines in ecological condition or the ineffectiveness of certain management actions. This requires internal leadership and the ability to negotiate bad news in the external environment of funders and policy makers.

7. Develop a monitoring framework with a hierarchy of monitoring to reflect different management aims

For land managers who are seeking to maximise biodiversity gains across multiple landscapes, development of a monitoring framework can assist in ensuring clarity of purpose and message, consistency in approach and maintenance of standards. This needs to recognise different scales of management and must have sufficient flexibility to enable adoption of the most relevant suite of tools for the local situation.

A hierarchy of monitoring recognises that, for park managers, there is a spectrum of management aims and that depending on the level of risk, these may require different types of information, ranging from systematic observation to highly quantitative monitoring. Ecological monitoring can mean different things to different stakeholders, both scientist and non-scientist. For example, is the purpose of the monitoring to inform park management decisions or to provide reporting on policy compliance? Does the monitoring seek to assess long-term change in resource condition or assess the effectiveness of specific management actions? Is the monitoring for surveillance of emerging issues or is it to determine if particular attributes of park condition fall within a desirable range?

Parks Victoria has categorised its ecological monitoring into several broad categories, based on their aims (Parks Victoria 2009). They include (1) systematic recording of *incidental observations*, (2) *systematic surveillance monitoring*, including rapid assessment and photo point monitoring, and (3) *protocol-based monitoring* based on scientifically robust methods using established protocols to produce quantitative data to test hypotheses and/or detect desired levels of change. With limited resources, surveillance-based monitoring can be a useful additional management tool to highlight both emerging issues and the need for further investment in more quantitative monitoring. It is important, however, that these different types of monitoring are not confused with each other.

Parks Victoria's 'Signs of Healthy Parks' monitoring framework recognises three levels of monitoring: *Activity monitoring* (the number and extent of management actions to protect a specified value or reduce a threat); *Effectiveness monitoring* (the extent to which the *activity* has reduced the threat); and *Environmental Outcome monitoring* (the resultant response of the targeted environmental value). The monitoring framework also should recognise that monitoring occurs across a range of management units. Depending on the question, this might be at the species, ecosystem, park or landscape scale. The type of units used will depend on management and ecological considerations as well as reporting requirements.

Recognising that resources for long-term monitoring will always be limited, a two-tiered model is being applied for its management effectiveness evaluation program. The first tier includes undertaking a qualitative but very systematic park manager assessment of more than 300 of its parks (approximately 90 per cent of the area of parks estate) every three years. This is based on the World Commission on Protected Areas Management Effectiveness Framework (see Hockings *et al.* 2005) and recognises a range of information sources from results of quantitative monitoring, to staff observation and experience. The second tier is to implement more targeted quantitative and protocol-based monitoring programs for a subset of parks and issues.

8. Develop a monitoring plan

One of the most useful tools for more effective monitoring programs is the development of a plan itself. A monitoring plan ensures that all parties are working on agreed goals, standards and designs. It brings together the elements of why (purpose and objectives), how and when (design and analysis), where (priority locations), and who (accountabilities). Some park agencies are well advanced in peer reviewed monitoring plans (e.g. US NPS Vital Signs program, see Stevens *et al.* 2005; MDBC 2008; Fancy *et al.* 2009). Parks Victoria has also developed a number of draft monitoring plans for its 'Signs of Healthy Parks' pilot program and proposes to have these for all of its high priority parks.

Investment in planning and design also includes the need to have well-developed systems to store, organise and enable sharing of monitoring data. While there are legitimate concerns about insufficient monitoring programs being implemented, there is perhaps an even stronger argument that data already collected should be more accessible for analysis and interpretation. With a wide range of individuals, government and non-government organisations, community groups and others collecting data across the parks system, the need for common data standards has never been greater.

9. Use available tools such as conceptual models and risk assessments to make more informed decisions about what to monitor

With many more potential monitoring questions than resources to support them, decision-support tools including environmental risk assessments and ecosystem conceptual models can assist in making more informed choices about priorities. Monitoring priorities and objectives should be based on the best available understanding of the relevant ecosystem drivers, priority values, threatening processes and interactions, and potential management responses.

In Parks Victoria's experience, this has included the development of a systematic environmental risk assessment process (see Carey *et al.* 2004), development of ecosystem conceptual models (White 2010), as well as the application of the Department of Sustainability and Environment's 'Actions For Biodiversity Conservation' threatened species management process. Land and water managers have been increasingly using ecosystem conceptual models to formally represent a summary of expert understanding about ecosystems (e.g. see Fancy *et al.* 2009). These models form the foundation for an integrated and holistic approach to management and research. Building on this work, Parks Victoria has recently developed strategic scale ecosystem models for each of the nine broad ecosystems across the parks estate (White 2010). These will be used as the foundation for management plans, development of conservation objectives and management and monitoring priorities. These strategic-scale models can form the basis for more refined localised or issue-based models. The further development and application of ecosystem models and risk assessments will increasingly be a major driver of priority setting for management responses and monitoring programs.

10. Build truly collaborative monitoring partnerships

Most park management agencies in Australia do not have all of the necessary skills and capacity to design and implement priority ecological monitoring programs. Effective monitoring partnerships are therefore not only desirable but critical to building sustainable programs to inform management needs. One example where Parks Victoria has created strong and open research partnerships with external science partners is through its Research Partners Program.

The adoption of a multi-faceted model for delivery of monitoring programs is the most effective model for Parks Victoria. This includes using our own staff where there are sufficient skills and capacity, using specialist ecologists from other agencies or its research partners, in some cases using fully contracted programs that require specialist skills (e.g. marine biodiversity monitoring) and finally supporting volunteer-based monitoring where possible. There are further opportunities for improving monitoring partnerships that need to be explored. There is a great need, and many exciting opportunities, for new collaborative monitoring partnerships involving the expertise and commitment of scientists, park managers, other government agencies, contractors, volunteers and students.

Collaborative monitoring partnerships will not be effective if they are perceived as a simple purchaser–provider arrangement. Our Research Partners Program has shown that most successful partnerships occur where park managers and science partners work with each other at each stage of the cycle, from developing and clarifying questions, to communicating outcomes. Monitoring questions can be improved if scientists understand park managers' management needs and recognise their often comprehensive local knowledge, while park managers can develop greater understanding of scientific methods and principles by working with scientists.

Successful monitoring partnerships require commitment. Most successful monitoring programs have been built on the persistence and energy of dedicated individuals and there is a need to build partnerships around these individuals to ensure that these programs can grow and be more effective in the longer term. The pressures within modern organisations to constantly reform and renew themselves work against sustained monitoring frameworks and programs. Continuity and persistence need to be actively argued for, especially during organisation reform periods.

Conclusion

The lessons in this chapter highlight that much thinking has been done across the world in assessing how best to use the complementary skills of land managers and scientists to develop well-targeted and management-focused monitoring. However the major task is now to implement this new thinking. This includes: (1) scientists working closely with Protected Area managers to promote the value of well-targeted management-focused monitoring, as a meaningful and practical tool to improve their own management performance; (2) commencing the job of filling obvious skill gaps and building more effective monitoring partnerships, including demonstrating case studies, based on common understanding of key questions and priorities; and (3) working with investors in biodiversity management and Protected Areas to demonstrate that targeted monitoring is both a useful and an essential requirement to quantify the conservation return on investment.

Acknowledgements

The author would like to acknowledge Dr Mark Antos, Dr John Wright, Dr Steffan Howe, Brian Doolan and Ian Walker for their valuable comments on this chapter.

Biography

Tony Varcoe is currently the Manager, Research and Management Effectiveness for Parks Victoria, a role he has held since 2004. He has worked in Victorian park management for 25 years across a wide range of roles from senior field-based management to policy development and park management planning. In his current role he is responsible for programs such as Parks Victoria's Research Partners Program, a collaborative program for management-focused research, and the 'State of the Parks' management effectiveness evaluation program. A member of the World Commission on Protected Areas, Tony is involved nationally in improving the use of evaluation and monitoring tools to inform adaptive management.

References

Biggs HC and Rogers KH (2003). An adaptive system to link science, monitoring, and management in practice. In *The Kruger Experience. Ecology and Management of Savanna Heterogeneity.* (Eds J du Toit, K Rogers and H Biggs) pp. 59–80. Island Press, Washington.

Carey JM, Burgman MA and Chee YE (2004). 'Risk assessment and the concept of ecosystem condition in park management'. Parks Victoria Technical Series No. 13.

Cheal DC, Westbrooke M, Gowans S and Gibson M (2007). Vegetation change in Victorian mallee parks. Unpublished report prepared for Parks Victoria by the Centre for Environmental Management, University of Ballarat, Victoria and the Arthur Rylah Institute for Environmental Research.

Cook CN, Hockings M and Carter RW (2009). Conservation in the dark? The information used to support management decisions. *Frontiers in Ecology and the Environment* **8**, 181–186.

Fancy SG, Gross JE and Carter SL (2009). Monitoring the condition of natural resources in US National Parks. *Environmental Monitoring and Assessment* **151**, 161–174.

Field S, O'Connor P, Tyre AJ and Possingham HP (2007). Making monitoring meaningful. *Austral Ecology* **32**, 485–491.

Hockings M, Leverington F and James R (2005). Evaluating management of protected areas. In *Protected Area Management: Principles and Practice.* 2nd Ed. (Eds G Worboys, M Lockwood and T De Lacy). Oxford University Press, Melbourne.

Lee W, McGlone M and Wright E (2005). 'Biodiversity inventory and monitoring: A review of national and international systems and a proposed framework for future biodiversity monitoring by the Department of Conservation'. Landcare Research Contract Report: LC0405/122. Department of Conservation, Wellington.

Legg CJ and Nagy L (2006). Why most conservation monitoring is, but need not be, a waste of time. *Journal of Environmental Management* **78,** 194–199.

Leverington F and Hockings M (2004). Evaluating the effectiveness of Protected Area management: the challenge of change. In *Securing Protected Areas in the Face of Global Change: Issues and Strategies.* (Eds C Barber, K Miller, R and M Boness) pp. 169–214. IUCN, Gland, Switzerland.

Lindenmayer DB (2009). Old forest, new perspectives: Insights from the Mountain Ash forests of the Central Highlands of Victoria, south-eastern Australia. *Forest Ecology and Management* **258**, 357–365.

Murray–Darling Basin Commission (MDBC) (2008). 'The Living Murray Icon Site condition report'. October 2008. MDBC Publication No. 58/08. Murray–Darling Basin Commission, Canberra.

Nichols JD and Williams BK (2006). Monitoring for conservation. *Trends in Ecology and Evolution* **21**, 668–673.

Parks Victoria (2007). 'Victoria's state of the parks report'. Parks Victoria, Melbourne.

Parks Victoria (2009). Environmental monitoring guide. Unpublished report. Parks Victoria, Melbourne.

Parrish JD, Braun DP and Unnasch RS (2003). Are we conserving what we say we are? Measuring ecological integrity within protected areas. *BioScience* **53**, 851–860.

Robley A, Wright J, Gormley A and Evans I (2008). 'Adaptive experimental management of foxes'. Final Report. Parks Victoria Technical Series No. 59. Parks Victoria, Melbourne.

Stevens SB, Milstead M and Entsminger AG (2005). 'Northeast Coastal and Barrier Network Vital Signs Monitoring Plan'. Technical Report NPS/NER/NRTR–2005/025. US National Park Service.

Victorian Auditor-General's Office (2010). 'Control of invasive animals and plants in Victoria's Parks'. Victorian Auditor-General's Report 2009–10, p. 23. May 2010. Victorian Auditor-General's Office, Melbourne.

Wahren CHA, Williams RJ and Papst WA (2001). Vegetation change and ecological processes in alpine and subalpine sphagnum bogs of the Bogong High Plains, Victoria. *Australia, Arctic, Antarctic, and Alpine Research* **33**, 357–368.

Woodley S (2010). Ecological integrity and Canada's national parks. *The George Wright Forum* **27**, 151–160.

White A (2010). Ecosystem models for Victorian ecosystems. Unpublished report for Parks Victoria.

12 MONITORING FOR IMPROVED BIODIVERSITY OUTCOMES IN THE PRIVATE CONSERVATION ESTATE: PERSPECTIVE FROM BUSH HERITAGE AUSTRALIA

Jim Radford, Murray Haseler, Sandy Gilmore, Angela Sanders, Adam Kerezsy, Max Tischler and Matthew Appleby

Lesson #1. Monitoring improves understanding of ecological variation.

Lesson #2. Awareness of Australia's biodiversity crisis depends on long-term monitoring.

Lesson #3. Monitoring has identified areas for conservation action and investment.

Lesson #4. There is a lack of institutionalised support for monitoring.

Lesson #5. Too often, the wrong thing is monitored in the wrong place at the wrong time.

Lesson #6. Monitoring often occurs without a defined question or purpose.

Lesson #7. Integrate monitoring into conservation management programs.

Lesson #8. Consider spatial and temporal scales for sustainable monitoring programs.

Lesson #9. Improve links between academia and the conservation sector.

Introduction

Independent not-for-profit conservation organisations, such as Bush Heritage Australia, Australian Wildlife Conservancy, Tasmanian Land Conservancy and Nature Foundation South Australia (SA), are purchasing and managing an increasing amount of land in Australia. These organisations share a common purpose in that their primary objective is biodiversity conservation. While they are considered part of the 'private' (i.e. non-government) conservation estate, they differ from private citizens who manage their land for biodiversity in that they rely on donations from individuals, philanthropists, foundations and corporations to fund their activities. Not-for-profit conservation organisations therefore, have an obligation to their supporters to address two fundamental questions: (1) How has acquiring the land improved biodiversity conservation (conservation return on investment)?; and (2) Is the land being managed in the best way possible for biodiversity conservation? Both of these questions require monitoring that is able to demonstrate change, or lack of change, in indicators of ecological condition in relation to management strategies over relevant time frames and spatial scales.

Since the beginning of privately funded conservation management, supporters and funders have always wanted to know how many baits were laid, how many kilometres of fencing were erected, or how many hectares of weeds were sprayed. However, the benefits of these activities for biodiversity have often been assumed without the outcomes being measured explicitly. Moreover, not-for-profit conservation organisations seldom had the expertise or resources to demonstrate the effectiveness of their management by addressing complex questions such as: 'What effect does your fire management have on grassland communities?' or 'Has removing stock improved the viability of the bird community?' Consequently, biodiversity monitoring tended to be an afterthought or was spasmodic, considered only after funds to cover 'core' operational activities were allocated. This is no longer the case. Conservation organisations are acutely aware that monitoring is required to demonstrate conservation return on investment and to inform adaptive management. The expectations of informed supporters, who are asking more sophisticated questions, have demanded greater emphasis on monitoring. Accordingly, the willingness and capacity of not-for-profit conservation organisations to meet this demand has grown, and these conservation organisations are often at the forefront of innovative monitoring programs. However, greater support and integration with academia and government programs is required to improve biodiversity monitoring across the sector. This chapter explores some of the key successes, deficiencies and requirements in biodiversity monitoring from the perspective of a national not-for-profit conservation organisation, specifically, Bush Heritage Australia.

Successes

1. Monitoring improves understanding of ecological variation

The effectiveness of a particular management action or strategy can be evaluated only within the context of the background ecological variation inherent in the system. Monitoring the spatial and temporal variation in populations provides the foundation for understanding how biota respond to environmental variation.

There are several long-term studies grounded in collecting fundamental ecological data that are increasingly valuable for understanding ecological variation across a range of environments, including rainforest (Graham et al. 2006; Williams and Middleton 2008), montane forest (Mackey et al. 2002), tropical woodlands (Woinarski 2004; Woinarski et al. 2010) and mallee (Ward and Paton 2004; Paton et al. 2005). These studies and others like them have increased our understanding of how biota respond to natural processes such as climatic variability, fire regimes and hydrological regimes, as well as anthropogenic processes such as pastoralism, habitat fragmentation, introduced species, altered fire and hydrological regimes, and potentially, climate change. By examining patterns and changes in abundance and distribution of biota across a landscape, monitoring provides the foundation for developing ecological concepts such as resilience, refugia, source-sink dynamics and metapopulations. These concepts are now key drivers for conservation organisations when planning their investment priorities and strategies.

A prime example of the link between fundamental ecological research and conservation management is the relationship between Bush Heritage and the Desert Ecology Research Group, led by Chris Dickman and colleagues from the University of Sydney (Dickman et al. 2001; Letnic and Dickman 2006; Dickman and Wardle, Chapter 18, this volume). Their research on small mammals, reptiles, birds and plants in western Queensland has grown from a single site in 1990 to a network of trapping grids and monitoring plots across multiple properties, two of which are now Bush Heritage conservation reserves while the others remain commercial pastoral properties. A key component of the research is a network of 13 weather stations that has enabled biotic responses to be related to climatic variation at a relatively fine grain of spatial resolution (especially for the arid zone). Over time, the research focus has developed from understanding natural patterns of variation in relation to long-term environmental conditions towards asking questions about the effectiveness of different management approaches. Ultimately, a thorough understanding of the relationship between key ecological drivers, conservation management strategies and ecological outcomes will result in better biodiversity conservation.

2. Awareness of Australia's biodiversity crisis depends on long-term monitoring

Ecologists are adept at describing and documenting the decline of species, the disintegration of ecosystems and the loss of ecological function from landscapes. This has been effective in providing quantitative evidence of biodiversity declines in Australia and raising the profile and awareness of the scale and magnitude of the problem.

There are numerous accounts of declining populations, species loss and dysfunctional landscapes that serve as testament to the mismanagement of Australian landscapes and underscore the problems we face to reverse biodiversity declines (Dickman *et al.* 2000; Ford *et al.* 2001; Woinarski *et al.* 2010). Such research underpins the case for increased investment to protect and restore biodiversity, and is the 'bread and butter' of not-for-profit conservation organisations seeking support.

Unfortunately, much of this research is correlative or based on space-for-time substitutions, and therefore limited in its ability to identify causal mechanisms of change; and often vague in articulation of solutions. However, the most compelling descriptions of biodiversity declines are often relatively modest in spatial scale but impressive in their longevity and durability. For example, Andrew Bennett from Deakin University has been surveying birds at a dozen sites in the box-ironbark forests of central Victoria about six times per year for over 14 years. The value of this monitoring is that it has been able to demonstrate long-term, gradual declines in numerous species (e.g. superb fairy-wren *Malurus cyaneus*, speckled warbler *Pyrrholaemus sagittatus*) and guilds (e.g. ground-foraging insectivores) (Mac Nally *et al.* 2009) that are not discernable at shorter time-scales typical of research funding cycles and government projects (3–5 years) (see Figure 12.1). It has also been able to show correlations between environmental variables (e.g. rainfall, flowering) and long-term fluctuations in functional guilds and particular species. An important lesson from studies like those of Bennett is that the consistent application of a particular method at the same sites over long time frames is likely to yield insights that are not possible with even the most elaborate snapshot studies.

3. Monitoring has identified areas for conservation action and investment

Deciding where to invest scarce conservation dollars is critical to the success of biodiversity conservation. Investment decisions informed by quantitative monitoring data are more likely to yield long-term benefits for conservation.

Bush Heritage Australia aims to conserve significant parts of Australia's most important *high conservation value land*: in other words, high quality habitat that is poorly represented in the national reserve system and under threat. Defining and assessing habitat quality is a constant challenge but nevertheless paramount if Bush Heritage (and similar organisations) are to make a significant contribution to biodiversity conservation. Bush Heritage uses many sources of information for assessing the conservation value of land but empirical monitoring data has been instrumental in several significant acquisitions. Most notably, the purchases of

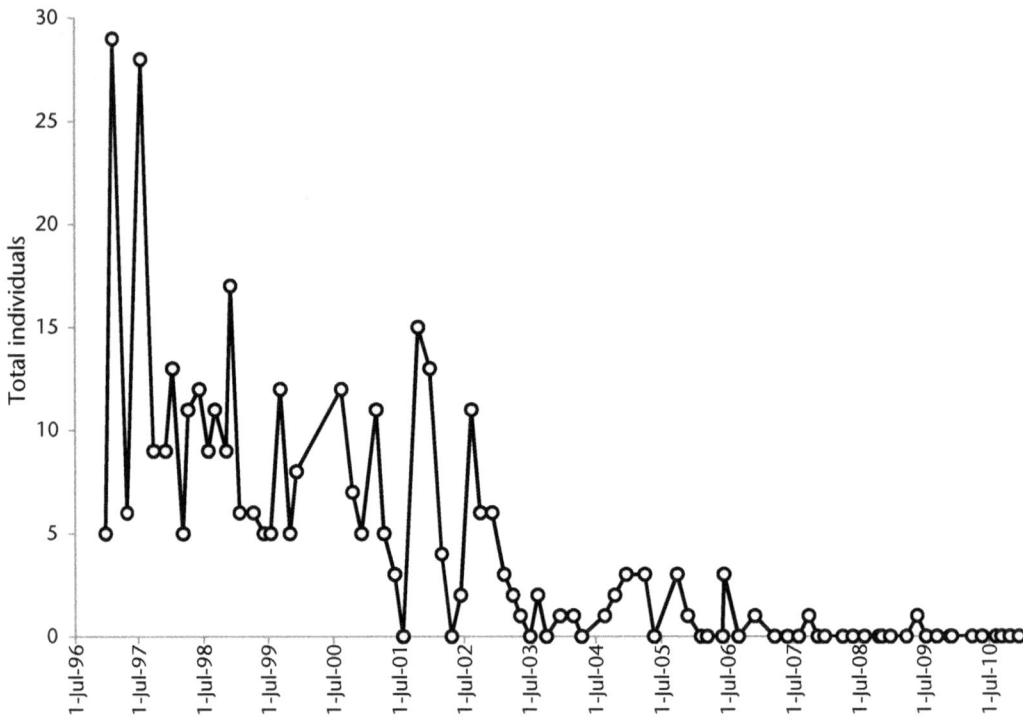

Figure 12.1. Total number of individuals of superb fairy-wren (*Malurus cyaneus*) detected on 12 monitoring sites in box-ironbark forests of central Victoria, 1996–2010. (Source: Andrew Bennett, Deakin University.)

Ethabuka Station (214 000 ha) in 2004 and Cravens Peak Station (232 000 ha) in 2005 were strongly influenced by the long-term monitoring and research by Chris Dickman and colleagues from the University of Sydney. Monitoring data were able to show that the diversity and abundance of small mammals and reptiles on these properties was of national significance, and demonstrate the importance of maintaining refuge habitat in the highly variable climate of the arid zone. This provided the necessary confidence to proceed with these ambitious purchases. These properties are now managed solely for biodiversity conservation, and with continued monitoring, the long-term conservation benefits are becoming apparent (Bush Heritage Australia 2010; Frank 2010).

Bush Heritage has recently developed and adopted a more sophisticated decision-support tool to assess conservation value that attempts to explicitly consider some key ecological processes that relate to land 'quality'. This framework, called Biodiversity Prediction using Ecological Processes, or BioPrEP, evaluates areas for investment based on seven biodiversity conservation goals, including capturing areas with relatively high primary productivity, protecting areas with high functional integrity, protecting viable populations of significant species or assemblages, and improving the level of protection of the least protected ecological types (Mackey *et al.* 2010). However, the utility of this framework is only as good as the data that support it: for example, our experience shows that more-informed investment decisions can be made when reliable and widespread species occurrence data can be incorporated into the analysis.

Failures

4. There is a lack of institutionalised support for monitoring

Governments, funding agencies, non-government organisations and regional catchment agencies have under-valued the importance of well-planned, long-term biodiversity outcomes monitoring. This has resulted in discontinuity and inconsistency in monitoring programs that compromise the value of the data collected.

Too often, biodiversity outcomes monitoring is piecemeal with specific monitoring activities tied to the life of a particular grant or project. Yet, the biodiversity benefits of projects will often not be apparent until some years after the actions are completed. Moreover, the effect of a particular project may, on its own, be marginal and very difficult to detect. However, the cumulative and interacting effects of many projects may be substantial but these effects may pass undetected in project-specific monitoring.

Without a coordinated and supported approach, biodiversity monitoring often suffers from inconsistent application of indicators and methods, temporal gaps in data collection, lack of data on key environmental drivers, lost data, and consequently, onset of a form of 'analysis-paralysis' in which it is perceived there is not enough information to analyse to generate robust findings (which may be true in many cases). Conversely, monitoring is often considered to be 'done' once the data are collected, without the commitment or planning to analyse and interpret the results.

5. Too often, the wrong thing is monitored in the wrong place at the wrong time

Selecting what to monitor, where to monitor it, how often to monitor it and when, is the crux of a good monitoring program. It is also the most contentious and challenging part of designing a monitoring program.

Selecting indicators requires a critical evaluation of the most appropriate ecological measure(s) at relevant spatial and temporal scales, as well as level of biological organisation (e.g. genes, species, populations, communities and ecosystems), within the context of the overall purpose of the monitoring program. Unfortunately, biodiversity monitoring often falls into the trap of 'counting what's countable' or has already been counted by someone else rather than 'counting what matters'. For example, biodiversity monitoring has an unhealthy reliance on broad remotely derived indices of 'landscape health' that are useful in their own right but, without validation and ground-truthing, contain little information on biodiversity *per se*. Similarly, biodiversity monitoring too often relies on single species monitoring, often of a threatened species, because of legal obligations under environmental protection legislation, resource constraints, funding appeal or the idiosyncratic interests of the investigator. This is often at the expense of more ecologically important indicators, such as community structure or ecosystem function, that are the fundamental ecological drivers of the system.

It is just as important to monitor in the right place, particularly if the objective of monitoring is to detect changes related to landscape-scale management. Pertinently, this may not necessarily be where the management actions are applied. Site selection requires an understanding of the hierarchical nature of landscape and catchment-level processes to ensure site-based measurements are taken where the impacts of management are likely to be manifest (Pringle *et al.* 2006). For example, in rangelands, sites located in relatively stable, intact areas away from areas of active incision or surface flows will not necessarily detect signs of recovery or dysfunction. Moreover, the monitoring results may be misleading if the intact areas are actually contracting due to wider landscape processes such as erosion and landscape desiccation.

6. Monitoring often occurs without a defined question or purpose

Monitoring without a defined question is just data collection. While there is merit in sentinel or background monitoring to describe patterns of habitat use in space and time, monitoring must have a defined purpose, and, ideally, be framed around specific questions (Lindenmayer and Likens 2010).

The perils of monitoring without a defined purpose include:

- neglect of key environmental or management co-variables;
- a tendency to default to preferred sampling methods or taxa without due consideration of appropriate indicators, sampling regime and statistical power;
- the sampling regime is time driven (i.e. at regular intervals) rather than process driven (i.e. after critical ecological events or life-history stages);
- the sampling regime (i.e. random, stratified, systematic, gradsect) is inappropriate; and
- insufficient thought is given to analysis and interpretation of data.

Solutions

7. Integrate monitoring into conservation management programs

Monitoring should be considered part of the core business of conservation management, with the design of the monitoring program driven by the objectives of the management strategies and interventions.

Biodiversity monitoring needs to be driven by, and designed to address, management effectiveness questions such as: 'Has feral predator control increased the population viability of native animal populations?' or 'What effect does fire management have on the structure and composition of plant communities?' Moreover, monitoring should provide information on how management strategies can be improved if they are not achieving the expected biodiversity outcomes. This requires definition of the current state and trajectory of the biodiversity measures, as well as an expected or desirable target condition so that progress towards the target can be tracked. Implicit in this is a consideration of alternative pathways to explain deviation away from the anticipated target. That is, why is the management intervention not having the desired effect? Monitoring should therefore imply decision rules: if condition x is not met in accordance with expectations (which may occur over decades), then alternative actions should be implemented or other explanations sought. More research is needed to identify thresholds for management interventions so that actions are not curtailed prematurely or prolonged unnecessarily.

Bush Heritage Australia has developed a monitoring program that explicitly focuses on ecological outcomes in the context of management actions and spatio-temporal variation in environmental drivers (Gilmore et al. 2008). Our monitoring is underpinned by two key elements: (1) a hierarchical framework composed of ecological goals (i.e. to maintain or restore: ecological function; viability of key species; functionally integrated communities; natural disturbance regimes; and ecosystem resilience), criteria for those goals, and indicators to measure the criteria that can be contextualised in relation to conservation values of the property; and (2) spatial representation of four stratification themes (i.e. environmental gradients; key conservation values; threats; and threat-abatement actions). Monitoring requires clear articulation of key conservation values; and goals, targets and indicators for each conservation value, which then determine the methods and sampling protocols used on a particular property.

8. Consider spatial and temporal scales for sustainable monitoring programs

To understand the hierarchical spatial structuring of ecosystems and processes, monitoring programs need to include indicators at different spatial scales. Similarly, biotic responses occur at different rates and vary across time, necessitating indicators on different temporal scales. However, monitoring programs rarely have the resources required to monitor all desirable indicators, so trade-offs between spatial and temporal coverage are inevitable.

When evaluating the trade-offs involved in designing a monitoring program, it is important to consider the sustainability of the effort required to maintain monitoring so that it can usefully inform management. That is, can we understand ecosystems, and species within them, through temporal replication at the expense of some spatial replication, or is it absolutely critical that all levels of the spatial hierarchy are adequately represented but perhaps monitored less often? Not-for-profit conservation organisations often monitor at relatively small spatial scales (e.g. a single reserve) but over comparatively long temporal scales, because they are not beholden to research funding or political cycles. A constant challenge, however, is to identify and monitor comparable 'control' sites that provide contrast in management activities and allow for increased rigour and confidence in monitoring results. Where controls are incorporated into monitoring programs, not-for-profit organisations are well placed to develop a good understanding of the land, its ecosystems and component species and processes, and therefore stand a better chance of turning monitoring results into useful on-ground decision-making.

9. Improve links between academia and the conservation sector

Closer links between academia and the not-for-profit conservation sector will result in more rigorous monitoring and better understanding of ecological concepts.

Not-for-profit conservation organisations have limited resources to expend on monitoring, despite the increasing pressure to demonstrate conservation return on investment. Closer links with academic researchers and students provides a win-win scenario. Researchers gain access to study sites representing a range of physical and biotic attributes, many with built-in manipulative management experiments. Land managers gain increased rigour in biodiversity monitoring, and access to intensively focused research effort around specific questions of interest. The partnership between the Desert Ecology Research Group and Bush Heritage Australia provides an example of the mutual benefits of such partnerships. However, we need to move beyond traditional funding sources such as Australian Research Council grants and student scholarships to seek new avenues for increasing resources and participation in monitoring. Collaborating with the corporate sector and other non-government organisations, such as Earthwatch and Birds Australia, may provide both a source of income and extra volunteer capacity.

Conclusion

The not-for-profit conservation sector now plays a significant role in biodiversity conservation in Australia, and this role is likely to increase. Monitoring the effectiveness of management activities is critical to maintaining donor support and investment, and improving conservation outcomes using an adaptive management framework. Through strengthening links with academic researchers and government agencies, not-for-profit organisations can also make a significant contribution to the way biodiversity monitoring is designed and conducted. In return, conservation organisations will continue to benefit from increased rigour in monitoring design, and therefore confidence in monitoring results. This translates to better decisions about where to invest, and how to manage the land, for biodiversity conservation.

Acknowledgements

We thank David Lindenmayer and Phil Gibbons for organising the workshop on Monitoring for Improved Biodiversity Outcomes, and the invitation to participate in the workshop. Chris Dickman and Andrew Bennett kindly provided data for this chapter, and their comments, along with those of Nicki Markus, improved earlier versions of this work.

Biographies

Jim Radford is the Science and Monitoring Manager at Bush Heritage Australia. He has a longstanding interest in biodiversity conservation in the temperate woodlands of southern Australia.

Murray Haseler is the Queensland Uplands and Brigalow Belt Ecologist at Bush Heritage. He has extensive experience with protected area management in Queensland, and has a particular interest in the ecology and conservation of mammals.

Sandy Gilmore is the South East Grassy Box Woodlands Ecologist at Bush Heritage. He is committed to deriving and applying ecologically relevant indicators of ecosystem health, particularly as applied to avian communities.

Angela Sanders is the Gondwana Link Ecologist at Bush Heritage. She was instrumental in the establishment of the Gondwana Link project, and is passionate about landscape restoration for biodiversity conservation.

Adam Kerezsy is Bush Heritage's Freshwater Ecologist. He has expertise in arid zone river systems, and all things fishy.

Max Tischler is the Gulf to Lake Eyre Ecologist at Bush Heritage. He has a fascination for the red sand country, having worked in the Simpson Desert for 14 years on a range of taxa including birds, mammals, reptiles and plants.

Matthew Appleby is the South West Botanical Province and Tasmanian Ecologist at Bush Heritage. He is a botanist with a specific interest in the ecology and conservation of native grasslands.

References

Bush Heritage Australia (2010). 'Bush Heritage Australia Annual Conservation Report 2009/10'. Bush Heritage Australia, Melbourne.

Dickman CR, Leung LKP and Van Dyck SM (2000). Status, ecological attributes and conservation of native rodents in Queensland. *Wildlife Research* **27**, 333–346.

Dickman CR, Haythornthwaite AS, McNaught GH, Mahon PS, Tamayo B and Letnic M (2001). Population dynamics of three species of dasyurid marsupials in arid central Australia: a 10-year study. *Wildlife Research* **28**, 493–506.

Ford HA, Barrett GW, Saunders DA and Recher HF (2001). Why have birds in the woodlands of Southern Australia declined? *Biological Conservation* **97**, 71–78.

Frank A (2010). Effects of cattle grazing on small vertebrates in the Simpson Desert, Australia. PhD thesis. University of Sydney, Sydney.

Gilmore S, Radford J, Pringle H and Foreman P. (2008). Ecological outcomes monitoring on Bush Heritage Australia reserves: evaluating real change in conservation outcomes. In *Australian Protected Areas Congress 2008 Proceedings: Protected Areas in the Century of Change*. (Eds I Garven and S Monk) pp. 42–44. Australian Protected Areas Congress, Queensland. <http://www.apac08.org.au/images/stories/apac%20papers%20web.pdf>

Graham CH, Moritz C and Williams SE (2006). Habitat history improves prediction of biodiversity in rainforest fauna. *Proceedings of the National Academy of Sciences of the USA* **103**, 632–636.

Letnic M and Dickman CR (2006). Boom means bust: interactions between the El Nino Southern Oscillation (ENSO), rainfall and the processes threatening mammal species in arid Australia. *Biodiversity and Conservation* **15**, 3847–3880.

Lindenmayer DB and Likens G (2010). *Effective Ecological Monitoring*. CSIRO Publishing, Melbourne.

Mac Nally R, Bennett AF, Thomson JR, Radford JQ, Unmack G, Horrocks G and Vesk PA (2009). Collapse of an avifauna: climate change appears to exacerbate habitat loss and degradation. *Diversity and Distributions* **15**, 720–730.

Mackey B, Lindenmayer DB, Gill M, McCarthy M and Lindesay J (2002). *Wildlife, Fire and Future Climate: A Forest Ecosystem Analysis*. CSIRO Publishing, Melbourne.

Mackey B, Gilmore S, Pringle H, Foreman P, van Bommel L, Berry S and Haseler M (2010). BioPrEP – a regional, process-based approach for assessment of land with high conservation value for Bush Heritage Australia. *Ecological Management & Restoration* **11**, 51–60.

Paton, DC, Rogers DJ, Ward MJ and Gates JA (2005). Birds and fire in the mallee heathlands of Ngarkat. *Wingspan* (Supplement) **15**, 21–24.

Pringle HJR, Watson IW and Tinley KL (2006). Landscape improvement, or ongoing degradation – reconciling apparent contradictions from the arid rangelands of Western Australia. *Landscape Ecology* **21**, 1267–1279.

Ward MJ and Paton DC (2004). Responses to fire of Slender-billed Thornbills, *Acanthiza iredalei hedleyi*, in Ngarkat Conservation Park, South Australia. I. Densities, group sizes, distribution and management issues. *Emu* **104**, 157–167.

Williams SE and Middleton J (2008). Climatic seasonality, resource bottlenecks, and abundance of rainforest birds: implications for global climate change. *Diversity and Distributions* **14**, 69–77.

Woinarski JCZ (2004). The forest fauna of the Northern Territory: knowledge, conservation and management. In *Conservation of Australia's Forest Fauna*. (Ed. D Lunney) pp. 36–55. Royal Zoological Society of New South Wales, Mosman.

Woinarski JCZ, Armstrong M, Brennan K, Fisher A, Griffiths AD, Hill B, Milne DJ, Palmer C, Ward S, Watson M, Winderlich S and Young S (2010). Monitoring indicates rapid and severe decline of native small mammals in Kakadu National Park, northern Australia. *Wildlife Research* **37**, 116–126.

13 PRACTICAL CHALLENGES IN MONITORING AND ADAPTING RESTORATION STRATEGIES AND ACTIONS

David Freudenberger

Lesson #1. There are different types of environmental monitoring for divergent objectives.

Lesson #2. Monitoring starts with eyeballs and a brain willing to learn.

Lesson #3. Monitor consequences of actions, not just environmental impacts.

Lesson #4. Monitoring frameworks, methods and systems are needed, but institutional culture and investment are more important.

Lesson #5. Common terminology is needed.

Lesson #6. Rewards and penalties are required.

Introduction

Greening Australia has been at the tough end of conservation for the past 30 years. We're in the business of land repair as a charitable-status environmental organisation with over 250 staff in every state and territory of Australia. Our mission is to facilitate the process of ecological restoration. We focus on native vegetation, because without it, there isn't much habitat for conserving Australia's extraordinary diversity of organisms from bacteria and beetles to bats and birds that seldom survive in highly altered landscapes of crops and non-native pastures. We work with a diversity of individuals, communities, businesses and governments to repair degraded landscapes that have been over-cleared, over-cropped, infested with weeds and feral mammals, or largely urbanised. We try to think and plan from landscapes to global scales, but in the end restoration occurs at the scale of individual seeds within a few millimetres of soil, one hectare at a time, and one landholder at a time.

We aim to be an innovative organisation, adapting our restoration strategies and tactics. We try to learn from our successes and failures. This has been tough to do in a systematic and efficient manner. The following are some of the lessons gained in building a learning and adaptive organisation. These lessons are based on many conversations across this organisation, as well as my own 25 years of experience as an ecological researcher inside and outside of Greening Australia.

Lessons

1. There are different types of environmental monitoring for divergent objectives

Too many monitoring projects lose their way due to a failure to precisely define their monitoring objective(s). The Australian Government's recently released biodiversity conservation strategy (Natural Resource Management Ministerial Council 2010) is a classic example of mixed monitoring objectives that are poorly defined. Two distinctly different kinds of monitoring are 'status' vs. 'strategy effectiveness' monitoring (CMP 2007; Salafsky *et al.* 2008). *Status* monitoring is about 'Evaluating the state of biodiversity …' using the language of Australia's *Biodiversity Conservation Strategy* (Natural Resource Management Ministerial Council 2010). *Status* monitoring should feed into State of the Environment reporting (e.g. Beeton *et al.* 2006). The excellent service provided by Birds Australia (www.birdaustralia.org.au) and its many thousands of citizen scientists is an important example of *status* monitoring of the distribution and relative abundance of Australia's birds which provides an insight into the state of a landscape and its ecological processes. A key objective of *status* monitoring is to provide early warnings of population declines, invasions or failure of ecological processes (e.g. vegetation condition; Gibbons and Freudenberger 2006). *Status* monitoring helps identify where active conservation interventions are required. Hence *status* monitoring needs to be long term and geographically widespread. The Wentworth Group of Concerned Scientists' (2008) proposal for a national environmental account is a worthy call for comprehensive and continental-scale *status* monitoring.

In contrast, *strategy effectiveness* monitoring is about questioning and testing what conservation interventions work once someone decides that the *status* of a particular population, species or other attribute of biodiversity warrants active intervention. Useful *strategy effectiveness* monitoring can be relatively small scale and over short periods of time, if the strategy is small scale and highly responsive like testing different techniques to restore grassy ground covers (e.g. Gibson-Roy *et al.* 2010). *Strategy effectiveness* is about intervention strategies and actions. In contrast, *status* monitoring is about intervention policies. The Natural Resource Management Ministerial Council (2010) biodiversity conservation strategy is a prime example of totally mixing up these two fundamentally different objectives for monitoring.

I see a third type, 'adaptive monitoring' (Lindenmayer and Likens 2010) as a framework particularly suited for long-term ecological field research (hypothesis testing). In this essay, I focus on the challenges of adaptively improving restoration actions (practices) through monitoring of *strategy effectiveness*. This is perhaps a subset of the broader *adaptive monitoring*.

2. Monitoring starts with eyeballs and a brain willing to learn

There is a commonly held assumption that monitoring to improve conservation strategies and tactics requires quadrats, tape measures and replicate plots. These are useful tools, but not essential. What is first needed is a burning desire by *individuals* to learn and improve. Such individuals are rare on the ground. We professional researchers too often assume others are addicted to exploring and learning. This is a weak assumption, rarely tested. Useful strategy monitoring requires a willingness to admit failure. Willingness to learn, innovate and change requires acknowledgement that 'best practice' isn't the best. Many people and their employers are threatened by the 'objective eye of monitoring'.

In contrast, I've worked with the likes of Justin Jonson who Greening Australia employed as a restoration officer in south-western Western Australia. Justin has the discomfort of never being satisfied with the status quo. When he started with us he had an informal look around at what previous restoration strategies (e.g. direct seeding) were delivering and he didn't like what he saw. Too often traditional direct seeding, in single widely spaced rows, results in an unnatural structure of densely vegetated rows with metres of bare and often eroding soil in between. Such informal monitoring, with a perceptive pair of eyes and a restless mind, resulted in an innovative package of restoration strategies and technologies that are delivering very different restoration outcomes than traditional direct seeding (Jonson 2010).

3. Monitor consequences of actions, not just environmental impacts

Andre Zerger, Judith Harvey and I were commissioned by the Australian Government to assess the biodiversity benefits of the Natural Heritage Trust (NHT) program. We took a case study approach where there had been considerable NHT investment (Freudenberger *et al.* 2004; Zerger *et al.* 2009). There was no point looking for broad environmental benefits because the quantum of NHT investment at the scale of Bioregions was miniscule (e.g. as little as $1.10/ha in highly stressed Interim Biogeographic Regionalisation of Australia (IBRA) regions within the Intensive Landuse Zone; Harvey and Freudenberger 2003). Through a network of contacts we identified a dozen project areas that had received multiple NHT grants, in some cases a few million dollars. When we took a closer look, clear objectives for each project were one of the most difficult, but critical pieces of information to find. It's nigh impossible to monitor and assess the conservation benefits of a project if you don't know the objectives.

The next critical step in monitoring conservation strategy effectiveness is knowing *who* did *what, when* and the *cost*. These four bits of *actions* data (see Figure 13.1) are the easiest to lose. The NHT funded the largest single revegetation project in Australia (600 ha) on one of Greening Australia's restoration properties in the Gondwana Link initiative in south-west WA (www.Gondwanalink.org.au). Sadly, records have been lost documenting what species of seed were used, in which paddocks, in which month and in which year. We can still monitor the very patchy *results* (see Figure 13.1) of this planting (e.g. what plants are alive), but we are finding it difficult to interpret the resulting patchy plant establishment since we don't know what was planted in a particular area, during a particular week.

Too often monitoring starts at the wrong end of things. Investors (individuals, communities and organisations) are naturally keen to know the *impact* (see Figure 13.1) of conservation strategies. But *impacts* are at the end of a critical sequence of events and processes or 'results chains' (CMP 2007). There is little value in monitoring *impacts* of revegetation if the *results* of such an *action* are dead plants and a bare field because the revegetation failed to establish. Too often monitoring programs are established to monitor the beneficial *impact* of a fox control program on small native mammals, rather than first monitoring the direct *results* of fox control (dead or live foxes). Again, there is little point in monitoring for *impacts,* if fox baiting only *resulted* in a 10 per cent decline in fox density.

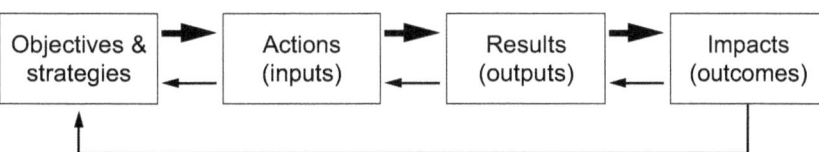

Figure 13.1. A simple logic framework that identifies the multiple points at which *strategy effectiveness* monitoring is required. Monitoring is needed for each box. Each thick arrow indicates the consequences of implementation. Each thin arrow indicates the critical interpretation and learning feedback loops.

4. Monitoring frameworks, methods, and systems are needed, but institutional culture and investment are more important

There are bookshelves of monitoring methods, great statistical designs, easy to use databases and slick data capture technologies, but they often collect dust. Some individual researchers, through sheer determination, and often at their own expense, have kept some long-term monitoring projects going (e.g. Likens 2004). However, whole organisation commitment to rigorous monitoring of conservation strategy *objectives*, *actions*, *results* and *impacts* (see Figure 13.1) is rare. It has to be both a top-down and a bottom-up commitment, from the Board and CEO to the staffer with a chainsaw and pack of herbicide. In Greening Australia's national office our Board and CEO committed to a whole of Federation monitoring system, invested tens of thousands of dollars in a customised web-based data capture system, but it never got traction in our regional offices. In contrast, our Capital Region office has slowly built up a comprehensive, project by project, database fully supported by field staff as well as senior managers. This database is now a high value tool for researching the effectiveness of landscape-scale restoration interventions. It is a spatially explicit record of nearly every revegetation and remnant protection project Greening Australia has been involved in over the past 20 years in the region (S. Wilson, pers. comm.). It has facilitated the research of five ANU Honours projects. Without it, the former New South Wales Department of Environment, Climate Change and Water and Greening Australia could not have assessed the variable conservation benefits of fencing woodland remnants rigorously stratified by landscape position (Briggs *et al.* 2008). This powerful tool is based on Greening Australia staff committed to rigorously recording their on-ground actions day-by-day over years.

5. Common terminology is needed

Monitoring methods and data management systems need to be based on common definitions for essential or core data fields. Atyeo and Thackway (2009) provide recommendations for the essential 18 core elements needed to capture key objectives, strategies, actions and results of on-ground conservation activities. Zerger *et al.* (2009) provide definitions for these core elements as well as a broader range of useful data fields. Salafsky *et al.* (2008) provide a 'standard lexicon' for classifying threats to biodiversity and conservation actions. This lexicon has been adopted by a wide range of international conservation organisations. Further development and adoption of such common language is critical for combining local, regional and national monitoring observations.

6. Rewards and penalties are required

Few of us like to fill out our annual tax return. We do it because we have to. Greening Australia, like all other incorporated organisations, goes through a monthly, quarterly and annual cost of financial accounting, reporting and auditing. How strange that we carefully monitor our finances with the assistance of a dedicated and talented team of accountants, but don't monitor our core business of restoration effectiveness with the same rigour and resources. It was no different during my 15 years in the Commonwealth Scientific and Industrial Research Organisation (CSIRO). Finances were monitored and reports carefully archived, but our core research, particularly our field data, was given no such care and attention. I still find it peculiar that research organisations spend so many resources managing and archiving secondary research outputs (e.g. journals and books), but no such care and resourcing dedicated to managing and conserving primary field data held in yellowing notebooks and now unreadable computer tapes (or it's all been chucked out!). Institutions involved in adaptive management, and funders of the necessary monitoring, must adopt the same discipline and even legal requirements that we commit to financial monitoring.

Conclusions

Adaptive conservation only occurs when individual leadership and passion are linked to an institutional culture committed to learning and improvement. Monitoring theory, systems, methods and technologies are needed, but their uptake will remain patchy, at best, until we as a society expect the same rigour of results-based reporting that we demand from our banks, public corporations and other urban institutions. We demand a return on investment from our human-generated capital, but not our investment in natural capital that supports us all. Hence environmental monitoring remains at the fringes of society rather than at the core like interest rates, Gross Domestic Product and employment figures.

Acknowledgements

Thanks to the many inside and outside Greening Australia for the numerous conversations on monitoring and adaptive restoration. Particular thanks to Susie Wilson and her colleagues for building a superlative project management system and culture.

Biography

David Freudenberger is Greening Australia's Chief Scientist. He has 25 years of ecological research experience, including 15 years with CSIRO. This experience includes research on wildlife nutrition, grazing management in rangelands, woodland restoration and monitoring frameworks. David's role in Greening Australia is to ensure the organisation's transformative landscape-scale initiatives are rigorously planned, implemented and evaluated based on the best available science. This includes leadership of Greening Australia's diverse range of restoration and carbon sequestration R&D projects.

References

Atyeo C and Thackway R (2009). *A Field Manual for Describing and Mapping Revegetation Activities in Australia*. Bureau of Rural Sciences, Canberra.

Beeton RJS, Buckley KI, Jones GJ, Morgan D, Reichelt RE and Trewin D (2006). 'Australia state of the environment'. Independent report to the Australian Government Minister for the Environment and Heritage. Department of the Environment and Heritage, Canberra. <http://www.environment.gov.au/soe/2006/publications/report/pubs/soe-2006-report.pdf>

Briggs SV, Taws NM, Seddon JA and Vanzella B (2008). Condition of fenced and unfenced remnant vegetation in inland catchments in south-eastern Australia. *Australian Journal of Botany* **56**, 590–599.

CMP (Conservation Measures Partnership) (2007). 'Open standards for the practice of conservation'. Version 2.0. Conservation Measures Partnership. <http://www.conservationmeasures.org/wp-content/uploads/2010/04/CMP_Open_Standards_Version_2.0.pdf> [Accessed 15 January 2011].

Freudenberger D, Harvey J and Drew A (2004). Predicting biodiversity benefits of the Saltshaker Project, Boorowa, NSW. *Ecological Management & Restoration* **5**, 5–14.

Gibbons P and Freudenberger D (2006). An overview of methods used to assess vegetation condition at the scale of the site. *Ecological Management & Restoration* **7** (S1), S10–S17.

Gibson-Roy P, Moore G, Delpratt J and Gardner J (2010). Expanding horizons for herbaceous ecosystem restoration: the Grassy Ground Cover Project. *Ecological Management & Restoration* **11**, 176–186.

Harvey J and Freudenberger D (2003). 'A spatial analysis of Commonwealth investment through the Natural Heritage Trust'. CSIRO Sustainable Ecosystems, Canberra.

Jonson J (2010). Ecological restoration of cleared agricultural land in Gondwana Link: lifting the bar at 'Peniup'. *Ecological Management & Restoration* **11**, 16–26.

Likens GE (2004). Some perspectives on long-term biogeochemical research from Hubbard Brook Ecosystem Study. *Ecology* **85**, 2355–2362.

Lindenmayer DB and Likens GE (2010). *Effective Ecological Monitoring*. CSIRO Publishing, Melbourne.

Natural Resource Management Ministerial Council (2010). 'Australia's Biodiversity Conservation Strategy 2010–2030'. Australian Government Department of Sustainability, Environment, Water, Population and Communities, Canberra. <http://www.environment.gov.au/biodiversity/strategy/index.html>

Salafsky N, Salzer D, Stattersfield AJ, Hilton-Taylor C, Neugarten R, Butchart SHM, Collen N, Master LL, O'Connor S and Wilkie D (2008). A standard lexicon for biodiversity conservation: Unified classification of threats and actions. *Conservation Biology* **22**, 897–911.

Wentworth Group of Concerned Scientists (2008). 'Accounting for nature. A model for building the National Environmental Accounts of Australia'. Wentworth Group of Concerned Scientists, Sydney.

Zerger A, Freudenberger D, Thackway R, Wall D and Cawsey M (2009). VegTrack – a structured vegetation restoration activity database. *Ecological Management & Restoration* **10**, 136–144.

14 MEASURING AND REPORTING ON CONSERVATION MANAGEMENT OUTCOMES

Sarah Legge and Atticus Fleming

Lesson #1. There's nothing wrong with champions.

Lesson #2. Disconnect between science and resource management.

Lesson #3. Measuring the wrong things.

Lesson #4. Lack of academic interest.

Lesson #5. Longevity.

Lesson #6. Make the scientists light a fire; make the land managers collect some data.

Lesson #7. Measure the right things – know your patch.

Lesson #8. Change the academic reward system.

Lesson #9. Promote the longer haul.

Introduction

Working for a non-government organisation presents a unique opportunity when it comes to monitoring. To a greater degree than other sectors, the link between our support and our performance is more direct. Our supporters expect accountability, both in terms of conservation results and cost-effectiveness. Added to this, our supporters are engaged, our objectives are their objectives, our agendas are aligned, and this makes the balance between striving for conservation outcomes and for other priorities relatively less complex compared with government conservation agencies.

Australian Wildlife Conservancy (AWC) currently owns and manages 22 sanctuaries across Australia, covering a combined area of over 2.6 million hectares, and capturing a wide variety of ecosystems – tropical Queensland rainforests, arid inland deserts, savannas of the Kimberley, tall forests of the south-west and the coastal habitats of the Gulf of Carpentaria. This variety means that a high proportion of Australia's mainland fauna are represented on AWC sanctuaries – over 65 per cent of mammal species, over 80 per cent of birds, and over 45 per cent of reptiles and frogs.

Our perspective on monitoring is influenced by having to design and execute (with colleagues) a national monitoring program for our sanctuaries. This program needed to add value to the process of conservation management by measuring and reporting on outcomes, rather than passively cataloguing change. To achieve this, we combined a top-down with a bottom-up approach, using ecological principles to construct the framework of the program, and populating the framework with site-specific detail based on an intimate knowledge of the biodiversity values and the threats to those values on each sanctuary. Our view is that a national scheme for monitoring Australia's biodiversity will need to include a healthy dose of 'bottom-up' detail in order to be meaningful and successful; yet this is the ingredient in the mix that is most easily omitted.

Successes

1. There's nothing wrong with champions

Most successful monitoring programs have been led by one or a few strong individuals (for examples, see Lindenmayer and Likens 2010). Paradoxically, this is often viewed as a failing, or a potential weakness, because the program is vulnerable to collapse when that person(s) leaves. But the alternative is to institutionalise monitoring, make it expensive, bureaucratic and intellectually and operationally unresponsive. Most would agree that's undesirable too, and advocate an unspecified and vague intermediate position between these two extremes. We take a different view: champions work, so we should foster them, and harness their abilities to contribute to a national monitoring program. There are some excellent analogues in the broader science community: the Max Planck Society in Germany blatantly recognises and celebrates the intellectual rigour, energy, and vision that a single individual can bring to a program of research, by explicitly structuring the Institutes and research programs around high calibre researchers.

Under a good champion, the scale, personnel and details of the monitoring program come and go, but the culture and commitment of the team and the endeavour outlives these changes. Having a good leader and a vibrant and enthusiastic team is an essential ingredient for ensuring longevity and consistency. Granted, there are issues with succession, but these are surmountable and trivial compared to an ineffective and pointless monitoring program.

AWC's challenge, to drive continued improvement in conservation effectiveness and promote accountability to supporters, was to design a nationally consistent monitoring system for a very diverse set of properties. The system also had to cope with all the usual obstacles of a long-term program – changing staff, resource constraints, insufficient knowledge, and so on. In AWC's monitoring system, the ecological health of a property is measured annually against a suite of indicators grouped into three themes (species, ecological processes and threats); the annual assessment includes up to 50 indicators, depending on the property. We have rolled out the system to about half our sanctuaries now; the longest sanctuary program is now in its eighth year.

Implementing and reporting on the ecological trends on our properties has been a nerve-racking undertaking. Our supporters want to see the numbers of bilbies and burrowing bettongs increasing, the numbers of feral animals decreasing, the size of fires reduced and nutrient retention in soils increased. If we get the management wrong, and the monitoring results are not in our favour, that ugly fact is there in black and white. Perhaps this risk is one of the reasons that risk-averse governments haven't invested in decent monitoring and reporting. The conundrum is that our biodiversity can't be saved by a straightforward application of a set of management rules because we don't yet know those rules, and so the probability of failure, at least initially, is high. It's a platitude that the only way we can reduce that probability is by allowing ourselves to make a few mistakes, but agreeing that those mistakes will only happen once because we'll be measuring the consequences of our management.

Failures

2. Disconnect between science and resource management

The lack of communication between managers and scientists has become dogma; the divide is conceptual, experiential, intellectual, practical, social and institutionalised. In the worst expression of this problem, a lack of respect for the perspective and skills of the other group is evident and the consequences are clear – scientists don't address the issues that managers need feedback on, and managers ignore, or are never even exposed to the advice from academics.

3. Measuring the wrong things

A small number of programs track changes (typically, a loss) in elements of biodiversity at regional or continental scales. For example, we can follow the loss of species at bioregional scales (Sattler and Creighton 2002; Burbidge *et al.* 2008), increases in the numbers of threatened species at state and national levels, and rates of land clearance (Australian Bureau of Statistics, National Land and Water Resources Audit, Native Vegetation Inventory Assessment). These measurements are useful for tracking biodiversity at the broadest scale and can lead to policy change, such as the changes to land-clearing legislation that were associated with monitoring of land-clearing rates. However, in isolation they are overly reductionist and have neither a social nor an economic price tag (Wentworth Group of Concerned Scientists 2008). In particular, they are untied to management and lack context, yet we desperately need to know which specific management actions slow or reverse rates of biodiversity decline. For this, we need site-based monitoring (of species, communities, processes and threats) that is carefully designed to measure management effectiveness.

Ironically, as the rhetoric about the importance of management-related monitoring increases, so the average quality of monitoring programs that are attached to natural resource management diminishes. Output measures like the length of new fencing, the number of poison baits delivered, the number of planning workshops held and the number of farmers and indigenous communities engaged are proliferating, but they contain no information about whether that investment resulted in benefits for biodiversity. To some extent, this retreat to collating information on outputs that can be easily gathered from diverse regions, landscapes and socio-cultural settings, reflects the challenge of integrating and summarising the achievements of varied resource management projects. These output metrics can be collected within the lifespan of a standard funding cycle, they are always positive (you can't build 'minus 5 km' of fence) and therefore always good news. But we are subverting the conservation dollar to a social 'feel-good' dollar, without confessing that currently we are failing to report on the return for investment in our natural capital. We are wasting money, and losing biodiversity.

4. Lack of academic interest

Increasingly, universities and other research institutions are neither leading, nor collaborating in research to help develop or support good monitoring. Even finding an appropriate and interested supervisor for a student working on an important question embedded within a monitoring program is becoming a challenge. With a very small number of notable exceptions (led by champions), research programs with a focus on monitoring and the collection of basic ecological data are not an attractive proposition for academics seeking grants, tenure and publications. On top of that, monitoring suffers from connotations of intellectual inferiority.

5. Longevity

Australia's poor record of long-term monitoring has very little to do with the resourcing available; consider the Atlas project of Birds Australia (Barrett *et al.* 2003) which is one of our most comprehensive long-term national data sets, and was collected by an army of volunteers. It does have something to do with the short life cycles of resourcing, which dissuade scientists and managers from embarking on data collection that requires multiple years for trends to become apparent. However, the lack of champions and vision is a more important impediment even than the resourcing. For example, some individuals have been remarkably successful at maintaining monitoring programs of longer duration than the average funding cycle; and similarly, some programs that are blindingly good candidates for longevity fail to operate as they should because no one has stepped up to lead and herald them.

Solutions

6. Make the scientists light a fire; make the land managers collect some data

We are all aware of the importance of melding science and management delivery; it's discussed at the highest levels. To effect this, though, managers and scientists need to work together, towards shared goals, at the ground level. They need to understand each other's perspectives. Hence the facetious sub-heading: a fire ecologist will benefit from a closer experience of how fires behave under different circumstances, and from understanding the operational constraints that the fire managers work within. Likewise, managers need to be involved in data collection and interpretation, in order to experience and appreciate the value of the adaptive management cycle. Integrating science and management requires pervasive social and cultural shifts within conservation agencies.

7. Measure the right things – know your patch

Technology, especially the various forms of remote sensing, allows us to monitor many ecological attributes and processes from the office desk. These approaches are enormously valuable, especially at regional, continental or even global scales. But if we want to know how species, populations and ecological processes are changing, we need repeated site-based field data.

A good monitoring program is based, of course, on asking the right questions, in the right way. However, this isn't just about good design – it's about spending real time on the ground, and learning about the systems and the communities that are being monitored, getting involved with the various stakeholders and understanding the local social and political constraints and opportunities. Knowing your patch is a dynamic process. The monitoring should be continually helping to improve your understanding, and the monitoring system needs to be flexible enough to incorporate the accumulating insights, and ruthless enough so that when parts of the program are not working out (i.e. not providing insight), they get dumped. The last two decades have seen a diminishing of investment in field ecology and everything that goes with supporting that, both from research institutions as well as government conservation agencies. If we want to ask the right questions, we need to reverse this shift and get our monitoring machinery (scientists and managers) back out on the ground.

8. Change the academic reward system

As the performance indicators for academics become increasingly one-dimensional, a scientist's relevance to society is devalued. If we want environmental science to avoid a slide towards benign indulgence, we need to diversify the reward system for academics, and give them points for contributing to social, economic and environmental good, points for investing effort in projects that are a longer, harder slog and points for carrying out work (like basic wildlife research) that will never make it into a high impact journal.

9. Promote the longer haul

There is no doubt we need to establish a more continuous funding model for biodiversity monitoring that is decoupled from policy and funding cycles. That doesn't mean the funding shouldn't be performance based. In fact, funding should be ruthlessly performance dependent, but milestones should be related to an ecologically sensible timetable, and the outcomes sought should be the ones that matter (so kilometres of fence built is inadequate, but the ability to measure and report on the population trend of a threatened species protected by the fence does matter).

Funding isn't the only issue discouraging longevity. The problems listed above – of the science–management disconnect, the attrition of field ecology, the stigma associated with monitoring (as opposed to 'real' science), the poor quality of many monitoring programs – all these factors lead to a loss of morale and faith in the discipline, and addressing them will promote longevity.

Longevity shouldn't mean years of blind faith; a well-designed monitoring system that is explicitly tied to management and is soundly question based will tend to generate insights and suggest further opportunities at relatively short as well as longer time scales. A great example of this is the fire pattern monitoring for Kruger National Park in South Africa, which stretches back six decades and spans distinctly different periods of fire management approaches (van Wilgen *et al.* 2004). In the short term, this monitoring provided feedback to fire managers about performance against their annual prescribed burning targets. At the decadal scale, the monitoring has provided invaluable insight into our ability to control specific aspects of a fire regime.

Conclusions

We need to remember that developing a good national monitoring program, and even turning this into national environmental accounting, isn't just about promoting cross-agency cooperation, standardising metrics and integrating the monitoring with policy (all of which is essential). It's also about fundamental changes on the ground level by conservation agencies and research intuitions. We need to make the discipline of monitoring relevant, high quality, influential, vibrant and appealing.

Perhaps we should invent a new term altogether. 'Monitoring' is loaded with a reputation of tedious repetition, unrealised goals and disenfranchised staff collecting data that may (or more likely, may not) come to anything useful at some indeterminate point. Often it defaults to the passive recording of seemingly inevitable declines. But monitoring for effective and adaptive conservation management is none of those things. It's intellectually very challenging. Understanding complex and varied natural systems is hard. It should be viewed as a biologist's ultimate fantasy, where you have the opportunity to figure out how all the bits in ecosystems fit together; it's infinitely more interesting than scoring the behaviour of beetles in plastic takeaway tubs in a dim lab. And finally, monitoring biodiversity is a precious responsibility and custodianship; we need to take it more seriously, and get better at it.

Acknowledgements

For significant influence, inspiration and discussion about conserving our biodiversity, we thank John Woinarski, Alaric Fisher, Chris Johnson, David Lindenmayer, Andrew Cockburn, Stephen Garnett, Henry Nix, Penny Olsen, and all our AWC colleagues, especially Matt Hayward, Manda Page, John Kanowski, Tony Fleming, James Smith, Katherine Tuft, Alex James and Steve Murphy. Thanks also to David Lindenmayer and Max Bourke for initiating this discussion and contribution.

Biographies

Sarah Legge leads the Conservation and Science Program for the Australian Wildlife Conservancy, which has included the development and implementation of a monitoring framework that reports on trends in the ecological health of AWC sanctuaries. The Conservation and Science Program is a broad research program in wildlife ecology, with a particular focus on its application to practical conservation problems. Before making the transition to applied conservation biology, Sarah's background was field research in evolutionary and behavioural ecology.

Atticus Fleming is the inaugural Chief Executive of AWC, which manages more than 2.7 million hectares round Australia. Previously, he worked as a policy advisor on the personal staff of Australia's longest serving Federal Environment Minister, playing a major role in biodiversity law reform as well as issues such as climate change and fisheries management. Before that, Atticus worked as a constitutional lawyer and a corporate lawyer.

References

Barrett G, Silcocks A, Barry S, Cunningham R and Poulter R (2003). *The New Atlas of Australian Birds*. Royal Australasian Ornithologists Union, Melbourne.

Burbidge AA, McKenzie NL, Brennan KEC, Woinarski JCZ, Dickman CR, Baynes A, Gordon G, Menkhorst PW and Robinson AC (2008). Conservation status and biogeography of Australia's terrestrial mammals. *Australian Journal of Zoology* **56**, 411–422.

Lindenmayer DB and Likens GE (2010). *Effective Ecological Monitoring*. CSIRO Publishing, Melbourne.

Sattler P and Creighton C (2002). 'Australian Terrestrial Biodiversity Assessment'. National Land and Water Resources Audit. Land and Water Australia, Canberra.

van Wilgen BW, Govender N, Biggs HC, Ntsala D and Funda XN (2004). Response of savanna fire regimes to changing fire-management policies in a large African National Park. *Conservation Biology* **18**, 1533–1540.

Wentworth Group of Concerned Scientists (2008). 'Accounting for nature. A model for building the National Environmental Accounts of Australia'. Wentworth Group of Concerned Scientists, Sydney.

15 MAKING MONITORING WORK FOR CONSERVATION: LESSONS FROM THE NATURE CONSERVANCY

Jensen R. Montambault and Craig Groves

Lesson #1. Involve local communities in monitoring to improve monitoring and management.

Lesson #2. Target data to help leverage pilot projects to multiple places.

Lesson #3. An iterative review process should be conducted by both peers and senior managers.

Lesson #4. Diversity in monitoring is good, but challenging.

Lesson #5. Monitoring must be designed to answer questions that managers want answered.

Lesson #6. Stop reinventing the wheel and learn from meta-analyses.

Lesson #7. Special funding can drive standards across multiple projects.

Lesson #8. Evaluate the most effective form of monitoring guidance.

Lesson #9. Senior management can drive standards across highest risk projects.

Introduction

The Nature Conservancy has been working in earnest for nearly a decade to improve its efforts to measure and monitor the effectiveness of its conservation work across the globe. As a founding member of the Conservation Measures Partnership (CMP) in 2004 (CMP, www.conservationmeasures.org), The Nature Conservancy was an early adopter of the *Open Standards for the Practice of Conservation* that includes standards for monitoring and adaptively managing projects (CMP 2007). At the same time, a grassroots group of conservation biologists and practitioners in The Nature Conservancy developed a systematic approach to strategic planning for conservation projects with accompanying desktop software that explicitly included tracking indicators through time to measure achievements against expectations. More recently, The Nature Conservancy launched a set of pilot projects for the purpose of improving project evaluation through monitoring plans. These plans were peer-reviewed in a series of two 'Measures Summits' attended by project staff, partners and senior managers, including The Nature Conservancy's chief executive officer (CEO). Needs identified at these summits led to working guidance papers on monitoring for field practitioners and developing a 'Measures Business Plan', created with field staff, endorsed by executive leadership and intended to mainstream monitoring and evaluation at The Nature Conservancy.

Despite these considerable efforts in The Nature Conservancy and other biodiversity conservation organisations, a recent survey of 29 organisations and funders (Muir 2010) suggests that making decisions based on monitoring remains far from a mainstream practice even among members of the CMP. Only about 10–30 per cent of ongoing work is guided by adaptive management and approximately 7 per cent of projects complete the adaptive management cycle. As reasons for this, respondents to the survey recite a litany of impediments, such as inadequate time, funding, staff, leadership and accountability. In spite of these challenges, there are signs in The Nature Conservancy, at least, that monitoring our work's effectiveness and managing based on this information has gained tangible traction in the last 2–3 years. Below, we summarise some of the most important lessons we have learned in the monitoring and adaptive management journey.

Successes

1. Involve local communities in monitoring to improve monitoring and management

The Nature Conservancy worked with partner communities in the Bismarck Sea (northern New Ireland Provence, Papua New Guinea) to establish marine protected areas (MPAs) supporting spawning aggregations of large, vulnerable reef fish species. Sites were selected based on the presence of independently verified spawning aggregations, the political will of the communities and their interest in taking responsibility for enforcement and monitoring. Preliminary results suggest that involving communities in both management and monitoring reinforces their understanding and commitment to long-term conservation results (Hamilton *et al.* 2011), building on previous assessments of monitoring and co-management in Pacific Islands and beyond (Moller *et al.* 2004; Carlsson and Berkes 2005).

This is an important lesson because scientists often debate the rigour and merit of so-called 'citizen science' efforts (Silvertown 2009). An important distinction in the case of these New Ireland MPAs is that while data are collected by individuals without formal scientific training, these people are much more than volunteer members of the public. They are the very people responsible for making decisions about, and enforcing the rules of, the MPA. In this case, an important step in adaptive management – sharing the results with managers – is actually accomplished through the biological and social science monitoring because the constituency is one and the same.

2. Target data to help leverage pilot projects to multiple places

The Sustainable Rivers Project is a collaboration between The Nature Conservancy and a United States (US) Federal Government agency (US Army Corps of Engineers) to alter the current flow regimes of Corps-operated dams in a way that will benefit freshwater ecosystems without detracting from other dam operations (e.g. hydropower generation, flood control, recreation). Starting with a single pilot project on the Green River, Kentucky (US), this effort has been leveraged or replicated to eight other dams in major river systems in the US with the potential to extend the project to over 600 dams owned and operated by the Corps (Richter *et al.* 2003). Targeted monitoring on the Green River and additional demonstration sites on the Savannah River (Georgia), Bill Williams River (Arizona), and Big Cyprus Creek (Texas) was critical to the success of leveraging this project and developing common indicators of ecosystem health attributable to dam re-operation (Konrad *et al.* 2011). These results can be used or adapted to help design the flow prescriptions, guide active adaptive management and monitor the results of re-operation of dams on other rivers.

3. An iterative review process should be conducted by both peers and senior managers

Biodiversity trends and socioeconomic impacts are important long-term results of conservation work. They are impractical for evaluating conservation effectiveness in the shorter time frames required by funders and senior managers (Black and Groombridge 2010). The Nature Conservancy has developed a new management practice that combines senior manager and peer-review to focus on both near-term results as well as longer-term conservation outcomes. In this process, the senior manager of all field programs expects all priority projects to create a business plan with a measures component including financial, institutional, as well as near and longer-term conservation benchmarks. These plans are reviewed semi-annually and measurable progress is used, in part, to allocate discretionary funding. In addition, project teams are supported through periodic peer-review workshops where teams within a similar theme (e.g. capacity-building for indigenous groups, long-term sustainable financing of protected areas) meet to learn from and critique each other's business plans and monitoring efforts.

Failures

4. Diversity in monitoring is good, but challenging

In projects or landscapes where multiple partner organisations are working together (e.g. Gondwana Link in south-western Australia), it can be challenging to ensure a coordinated and consistent approach to monitoring. This is especially acute where organisations have nation-wide monitoring protocols that they need to follow. The Nature Conservancy's Australia Program works entirely with partners, most of whom have their own monitoring protocols (at an organisational and project level), to deliver conservation results. These different monitoring approaches are often necessary given the variability across projects. At the same time, this diversity may be a disadvantage in interpreting results across a range of sites monitored by different partners and in reporting results to donors.

5. Monitoring must be designed to answer questions that managers want answered

Monitoring of conservation efforts is often conducted in collaboration with universities because of the expertise they have and to save the conservation organisation time and money in these assessments. But are these results always useful? The Nature Conservancy surveyed 16 tallgrass prairie restoration projects in the central US to evaluate how the effectiveness of restoration was being monitored and if the results were being used. While most (14 of the 16) projects conducted some kind of biological monitoring, only those projects (9 of the 16) collaborating with a university produced any publications. In general, fewer than 40 per cent of peer-reviewed conservation-oriented articles are relevant to management (Fazey *et al.* 2005). In this case The Nature Conservancy concluded that research conducted on its grasslands sites by universities was often basic research that had no direct management implications, or was not presented in venues and formats that are accessible to managers. Another important finding was that 15 of these projects had zero staff time or funding available for conducting analyses, suggesting that even when managers allocated resources for monitoring, there was little thought given to follow through on the final adaptive management steps of analysing and using the results. Biological monitoring must be designed with management needs in mind in order to be more than an academic exercise.

6. Stop reinventing the wheel and learn from meta-analyses

Meta-analysis involves synthesising information on how a conservation intervention works across a wide range of sites to determine under what conditions it is most effective. These synthetic reviews can be very useful to conservation managers (Seavy and Howell 2010). The Nature Conservancy, with its hundreds of field sites and staff, should be poised to contribute significant meta-analyses to guide future conservation efforts. Yet, to date The Nature Conservancy has completed only one such evaluation on the effectiveness of conservation easements (Kiesecker *et al.* 2007). One explanation for this meagre result is that diversity in monitoring (Lesson #4) often means that data collected across multiple sites are incompatible or lack the necessary formal *a priori* commitments amongst partners (e.g. memorandums of understanding, dedicated data and communications coordinator) to share data. Beyond doing its own meta-analyses, Conservancy field staff need to take better advantage of the growing number of published meta-analyses that inform commonly used strategies such as marine protected areas (Selig and Bruno 2010) as well as websites such as the Collaboration for Environmental Evidence (http://www.environmentalevidence.org/) and the Centre for Evidence-Based Conservation (http://www.cebc.bangor.ac.uk/) which also summarise and document the effectiveness of a variety of conservation interventions.

Solutions

7. Special funding can drive standards across multiple projects

The Nature Conservancy, and a group of nearly 50 partners, has collaborated to fund a monitoring program across multiple sites and projects involved with Skagit River Basin (North American Pacific Northwest) estuary restoration. The Nature Conservancy facilitated a conversation among partners that identified the lack of regional predictive models for ecosystem processes (e.g. sedimentation flow) as a need that partners could both provide data for, and use the results from, to improve their restoration efforts. Working with major foundations and government funders, this group created an action fund that all partners can apply to for their routine conservation and monitoring work, with special funding reserved to provide extra money to organisations conducting additional monitoring to provide data needed for these regional predictive models. The Nature Conservancy is creating a similar supplementary monitoring fund to evaluate conditions for success on the Sustainable Rivers Project (Lesson #2). In both cases, we hope the funding will provide both incentives and a capacity to implement monitoring that can be used in a meta-analysis (Lesson #6) and will help solve some of the challenges presented by data collected by multiple agencies (Lesson #4).

8. Evaluate the most effective form of monitoring guidance

The Nature Conservancy surveyed approximately 140 staff members who are specifically trained to guide project teams through the adaptive management process (called conservation action planning coaches) to understand what kind of guidance and training they would like to receive to improve monitoring of conservation work. The 73 coaches who responded strongly preferred self-paced guidance in the form of self-paced online training (www.conservation-training.org), recordings of live internet-based lectures and discussions, and downloadable guidance documents and case studies over more personnel-intensive training, such as in-person workshops and a telephone- or email-staffed help-desk. This is important because in-person training and other assistance have been traditional approaches in The Nature Conservancy and are intensive investments in staff time and travel costs. Even among the self-paced forms of guidance, there is a wide range of resource investment that could be dedicated to any product and we need to carefully evaluate how and how often these products are accessed. This includes soliciting pre-publication review from experts and target audiences and comparing different methods of publicising monitoring guidance as well as tracking how these resources are used.

9. Senior management can drive standards across highest risk projects

In addition to guidance that can be freely accessed by anyone in the conservation community, particular high-risk projects merit special attention from senior management. These projects may be considered high risk because they are unusually sensitive from an ecological perspective, very expensive, generate a great deal of press, their strategies have a high degree of uncertainty, or they are currently or are likely to be replicated widely (Hummel *et al.* 2009). In such cases, it is critical that the institution understands how the project is working and why.

The Nature Conservancy has addressed this issue by establishing a Project Review Committee that screens high-risk projects (defined as any contractual commitment over US$1 million) and makes recommendations to the President-CEO and Board of Directors on if (and how) these projects should proceed. This committee was founded to allay legal concerns over financial commitments, so it tends to focus largely on the financial aspects of a given project. In the past several years, the committee has expanded its purview to include ecological and social criteria more closely, especially as The Nature Conservancy has expanded its international programs. While there is room for improvement, this kind of intense scrutiny and delegation of board authority has led to improved planning and implementation of monitoring for expensive projects such as the 'International Paper deal' (US$231 million to purchase land from the timber company across 10 states in the south-eastern US in 2005) and the 'Montana Legacy Project' (US$0.5 billion to purchase ~125 000 ha of land from a timber company in 2009). We hope this kind of oversight will lead to an improved monitoring consciousness that trickles down throughout the organisation.

Conclusions

Substantial progress has been made in the last three years towards integrating monitoring and adaptive management into The Nature Conservancy's normal way of doing business (see Figure 15.1). This positive momentum can be attributed to several factors. First, we have made progress in changing staff perception that monitoring and evaluating our work is largely a scientific exercise to an organisational discussion that views monitoring as addressing questions that managers need answered. Second, we have realised that The Nature Conservancy cannot invest in publication-quality monitoring programs for all of its projects. We are focusing our investment in monitoring on those projects that have the greatest opportunity for learning and leverage as well as those that present some of the greatest risks to the organisation. In these cases, we need a higher degree of confidence that our strategies and actions are being effective, so we invest in monitoring designs with a higher level of inference. Finally, the best ingredient for success in advancing monitoring and evaluation applied to adaptive management throughout The Nature Conservancy (or any organisation) is having the most senior managers of field programs ask a simple question: What progress are we making against what we expect to do in our most important projects and how do we know?

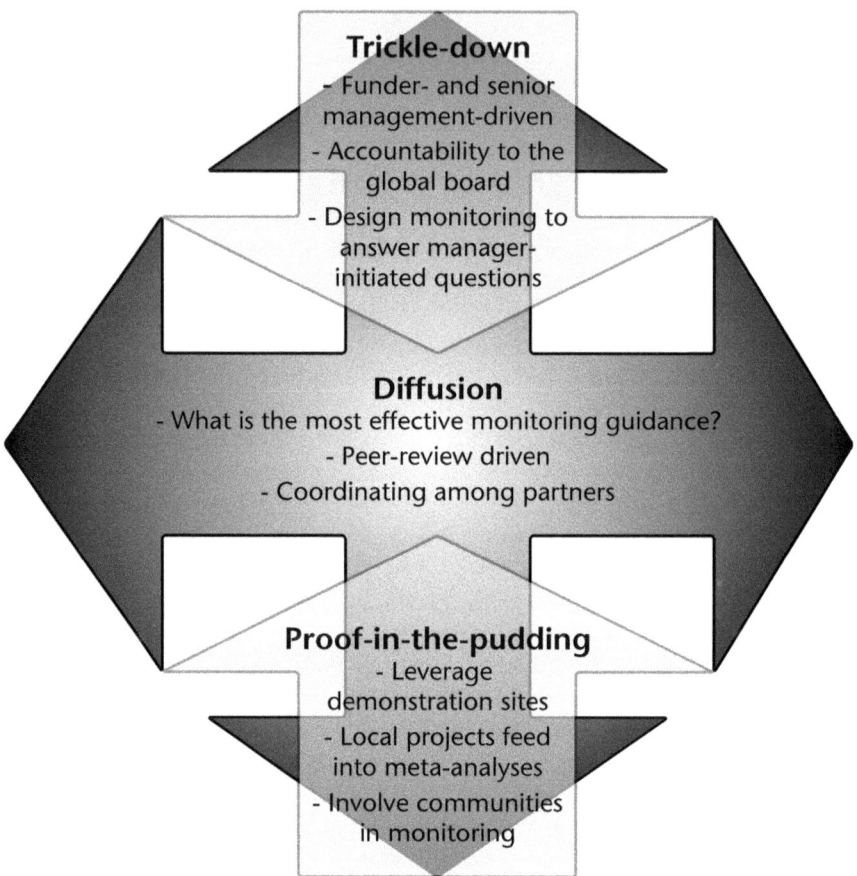

Figure 15.1. Lessons from The Nature Conservancy's efforts to incorporate monitoring and adaptive management into daily conservation work can be divided into three categories by which level of the organisation provides the impetus. A top-down or 'trickle down' approach means that monitoring originates with a management mandate. A bottom-up or 'proof in the pudding' approach means monitoring is initiated at localised sites to address specific needs and data eventually serves a wider audience. In the 'diffusion' model, different levels of employees, sites, projects, and organisations share lessons learned from monitoring across all levels. This model illustrates the need for simultaneous process rather than preferring one approach over another.

Acknowledgements

We thank R. Hamilton, J. Higgins, C. Konrad, J. Hardesty, J. Fitzsimmons, J. Fargione, H. Rowe, R. Fuller, J. Young, J. Ward and B. Runnels for contributing information about The Nature Conservancy examples in each lesson.

Biographies

Jensen Montambault joined The Nature Conservancy's Central Science division in 2008 helping conservation projects around the world assess the effectiveness of their work through appropriate monitoring designs, analysing data and incorporating results into management. Prior to joining The Nature Conservancy, Jensen worked for 13 years on conservation projects in Latin America, Caribbean, Africa, South Pacific and the US. She was a community environmental promoter in Peace Corps-Nicaragua, managed grants for the National Fish and Wildlife Foundation's USAID Neotropical Migratory Bird Conservation Program, and coordinated Conservation International's Rapid Assessment Program. She received her PhD from the University of Florida's Interdisciplinary Ecology program.

Craig Groves is the Director of Conservation Methods at The Nature Conservancy. His career with The Nature Conservancy started in 1984 as Director/Zoologist for the Idaho Natural Heritage Program. His science positions with The Nature Conservancy have included state, regional, and global levels as well as Conservation Science's Director of Conservation Planning from 1997–2002. Craig Groves also served as a Nongame and Endangered Species Biologist for the Idaho Department of Fish and Game and a conservation biologist and planner for the Wildlife Conservation Society. In 2003 he published *Drafting a Conservation Blueprint: A Practitioner's Guide to Planning for Biodiversity* through Island Press.

References

Black S and Groombridge J (2010). Use of a business excellence model to improve conservation programs. *Conservation Biology* **24**, 1448–1458.

Carlsson L and Berkes F (2005). Co-management: concepts and methodological implications. *Journal of Environmental Management* **75**, 65–76.

Conservation Measures Partnership (CMP) (2007). Open standards for the practice of conservation, Version 2.0. <http://www.conservationmeasures.org/wp-content/uploads/2010/04/CMP_Open_Standards_Version_2.0.pdf> [Accessed 11 February 2011].

Fazey I, Fischer J and Lindenmayer DB (2005). What do conservation biologists publish? *Biological Conservation* **124**, 63–73.

Hamilton R, Potuku T and Montambault JR (2011). Community-based conservation results in the recovery of reef fish spawning aggregations in the Coral Triangle. *Biological Conservation* **144**, 1850–1858.

Hummel S, Donovan GH, Spies TA and Hemstrom MA (2009). Conserving biodiversity using risk management: hoax or hope? *Frontiers in Ecology and the Environment* **7**, 103–109.

Kiesecker JM, Comendant T, Grandmason T, Gray E, Hall C, Hilsenbeck R, Kareiva P, Lozier L, Naehu P, Rissman A, Shaw MR and Zankel M (2007). Conservation easements in context: a quantitative analysis of their use by The Nature Conservancy. *Frontiers in Ecology and the Environment* **5**, 125–130.

Konrad CP, Warner A and Higgins J (2011). Evaluating dam re-operation for freshwater conservation in the Sustainable Rivers Project. *River Research and Applications*, Online Early, DOI: 10.1002/rra.1524.

Moller H, Berkes F, Lyver PO and Kislalioglu M (2004). Combining science and traditional ecological knowledge: monitoring populations for co-management. *Ecology and Society* **9**, 2.

Muir MJ (2010). Are we measuring conservation effectiveness? A survey of current results-based management practices in the conservation community. Conservation Measures Partnership. <http://www.conservationmeasures.org/measures-summit> [Accessed 11 February 2011].

Richter BD, Mathews R, Harrison DL and Wigington R (2003). Ecologically sustainable water management: managing river flows for ecological integrity. *Ecological Applications* **13**, 206–224.

Seavy NE and Howell CA (2010). How can we improve information delivery to support conservation and restoration decisions? *Biodiversity and Conservation* **19**, 1261–1267.

Selig ER and Bruno JF (2010). A global analysis of the effectiveness of Marine Protected Areas in preventing coral loss. *PLoS One* **5**, e9278.

Silvertown J (2009). A new dawn for citizen science. *Trends in Ecology and Evolution* **24**, 467–471.

16 BIODIVERSITY MONITORING FROM A COMMUNITY ORGANISATION PERSPECTIVE

Doug Robinson, Lisa Smallbone and James O'Connor

Lesson #1. Development of simple, standard programs for community monitoring is invaluable.

Lesson #2. Longevity is essential.

Lesson #3. Clearly defined objectives are essential.

Lesson #4. Adaptive management needs to become a structural component of monitoring programs.

Lesson #5. More long-term, scientific support is required to assist ecological monitoring by community groups.

Lesson #6. Lack of large, standard data sets reduces our capacity to monitor and evaluate national trends.

Lesson #7. Develop our monitoring resource base by investing in citizen science.

Lesson #8. Develop agreed ecological indicators and monitoring methods for a national monitoring program for ecosystems and species.

Lesson #9. Establish an ecological monitoring research program to support Natural Resource Management (NRM) community projects.

Introduction

Voluntary commitments by thousands of individuals, landholders and community organisations contribute significantly to biodiversity conservation in Australia (Curtis *et al.* 1999; Fitzsimons and Wescott 2001; Weston *et al.* 2003). These networks of volunteers represent an important form of citizen science and citizen conservation that underpins major NRM programs in Australia, including the National Reserve System program, the Caring for our Country (CfoC) initiative, and other programs in river health, sustainable land management and threatened species recovery.

Trust for Nature and Birds Australia are both conservation organisations that undertake biodiversity conservation actions in their own right, and also encourage and support thousands of individuals to become citizen scientists and conservationists. The primary purpose of this paper is to highlight some issues for the translation of monitoring policy into monitoring practice from the perspective of community organisations participating in biodiversity conservation in Australia.

Lessons

1. Development of simple, standard programs for community monitoring is invaluable

Some of the most effective NRM monitoring programs in Australia, such as Birds Australia's Bird Atlas project (Barrett *et al.* 2003) and Waterwatch Australia (Ryan 2003), rely largely on volunteers collecting data in a simple but standard way at multiple sites. Part of the success of each of these programs has been their effectiveness in motivating and supporting volunteers to participate in data collection as a contribution to knowledge and conservation. Their effectiveness has resulted from the development of detailed educational and extension resources that are provided to all volunteers, as well as regular feedback to the volunteers about the programs' results. Using these community-based monitoring techniques, both programs have contributed substantially towards biodiversity conservation and river health programs at national, state, regional and local scales.

For example, Birds Australia's Shorebirds 2020 project has been running since 2006, drawing on data collected by committed volunteers at key sites since 1981. It has successfully integrated a large number of regionally or locally based site monitoring efforts by clarifying site boundaries, standardising monitoring methods, determining the monitoring effort required to detect 'national' trends and investing significantly in training shorebird watchers in species identification and count methods. It has increased the number of regularly monitored sites from around 30 in 2004 to over 150 in 2010, with data collected by over 1000 volunteers. In line with the program's objective of being able to discern population trends for shorebirds, trend data are beginning to emerge which are forming the policy basis for a conservation advocacy tool to protect migratory shorebirds.

2. Longevity is essential

Longevity of any monitoring program is essential to determine ecological responses. As the Bird Atlas project illustrates so well, long-term data provide the potential to address many ecological questions; Bird Atlas data, for example, are increasingly being used to discern trends in abundance, distribution, responses to habitat change and responses to climate change by Australian birds (Olsen 2008).

Long-term monitoring data also provide opportunities to address questions relating to management interventions. In northern Victoria, we re-examined population data collected for the Grey-crowned Babbler (*Pomatostomus temporalis*) in 1995 and 2008 and evaluated changes in occurrence and group size in relation to habitat change and management interventions (Wilson *et al.* 2009). Results from that study showed that habitat interventions have had a significant, positive effect on babbler occurrence and group size (Wilson *et al.* 2009), subsequently helping land managers to refine their conservation guidelines and investment criteria for this endangered species.

3. Clearly defined objectives are essential

Monitoring activities need to be defined at the start of a project within a broader planning context that articulates the project's objectives, the strategies proposed to achieve those objectives, the specific actions contributing to implementation of the strategies and the assumptions underlying the strategies (Conservation Measures Partnership 2007).

The Australian Government has helped to address the issue of linking monitoring strategies to project objectives by formally articulating the links between a project's activities (outputs) and the Government's longer-term NRM outcomes as part of its CfoC initiative (Australian Government 2009). It has also developed extensive resources to assist groups prepare monitoring and evaluation plans as part of their project planning (e.g. Roughley 2009). However, most projects currently funded under CfoC or other government funds have monitoring, evaluation, reporting and improvement (MERI) plans geared to reporting on project milestones and outputs, rather than to explicit biological or environmental condition targets linked to the project's environmental objectives.

Where NRM programs have been developed with specific environmental objectives, targets and indicators, the development and implementation of the conservation monitoring and evaluation process becomes clear (Conservation Measures Partnership 2007). For example, the environmental management plans prepared for The Living Murray Icon Sites represent an excellent example of a conservation program in which the environmental objectives of the project are clearly defined and the monitoring is closely linked to those objectives, specific environmental criteria and specific ecological indicators (MDBC 2006).

4. Adaptive management needs to become a structural component of monitoring programs

Given the biodiversity crisis in Australia, we must continue taking action to counter threats to biodiversity and improve the ecological condition of ecosystems and species. It is essential that we systematically learn from our interventions to improve the effectiveness of our conservation responses.

One of the recognised strengths of the English agri-environment program has been the strongly integrated research and monitoring program, based on a cycle of trend monitoring, diagnosis of causes of decline, testing of solutions, implementation and monitoring (Bradbury *et al.* 2004; Grice *et al.* 2004). Australia does not have such a well-developed research program linked to NRM programs (Gibbons *et al.* 2008). Facilitating this type of systematic, evidence-based approach to natural resource management could effectively improve biodiversity condition by testing and evaluating practical solutions for improvements in ecosystem condition habitat management (Bradbury *et al.* 2004) and threatened species management (Lunt *et al.* 2005; Wilson *et al.* 2009).

5. More long-term, scientific support is required to assist ecological monitoring by community groups

Every year, hundreds of community groups and community organisations receive public funding to implement conservation projects, a requirement of which is that those groups monitor and evaluate the results of their activities. Rightfully, the funding bodies want community groups to collect some information related to the project's objectives that will allow the investors to evaluate the success of the project in terms of achieving its environmental goals. However, many of the questions that need to be answered through monitoring programs are questions that require professional statistical expertise in designing studies to demonstrate the effectiveness of on-ground works (Watson and Novelly 2004).

An additional difficulty faced by most community organisations undertaking NRM projects with public funds is that their funding is for only 1–3 years but the ecological responses predicted as a part of the NRM project may require far longer to transpire (Watson and Novelly 2004; Australian Government 2009).

As noted previously, the Australian Government has provided extensive resources to assist with the formal development of monitoring programs in a project sense (Lesson #3). Where there is still a gap in monitoring programs for community projects is in the provision of resources and guidelines relating to the 'nuts ands bolts' of establishing an ecological monitoring program. In particular, ecological monitoring by community organisations could be improved by developing educational resources that set out basic scientific principles for any ecological monitoring program, including advice on the design of the monitoring program, advice on possible field methods and advice about data collection, analysis and storage. Providing groups with lists of additional resources and institutions that could assist them with monitoring would also be beneficial.

6. Lack of large, standard data sets reduces our capacity to monitor and evaluate national trends

With notable exceptions, such as Birds Australia's Bird Atlas and the Bureau of Meteorology's climate data, there are few consistent environmental data sets across regional and state/territory boundaries in Australia, limiting our capacity to evaluate trends in environmental condition or responses to management interventions and environmental change. Where such data are available, it has been possible to discern important trends that can influence subsequent policy and implementation (e.g. Grice *et al.* 2004; Olsen 2008).

7. Develop our monitoring resource base by investing in citizen science

Lack of sufficient baseline data for almost all classes of Australia's biological assets is a widely acknowledged deficiency in our attempts to track environmental condition (Wentworth Group of Concerned Scientists 2008; Natural Resource Management Ministerial Council 2010). Citizen science-based monitoring programs offer real opportunities to improve our knowledge of biodiversity and capacity to track trends for environmental assets.

Birds Australia's Atlas program demonstrates that encouraging slight modifications to existing recreational activities (i.e. birdwatching), and providing a simple and accessible framework for data collection, can generate useful data at minimal cost. The cost of gathering equivalent-value data has been conservatively estimated at over two million dollars per annum. While this program may not be perfect in terms of design and data quality, the eight million records generated to date by the Bird Atlas represent a significant baseline – one of very few collected at a national scale, and one which will only improve with the embedding of more structured programs and integration of data sets. Other citizen science programs such as Reef Watch and Waterwatch similarly demonstrate the power of people in generating useful, robust data for monitoring programs.

8. Develop agreed ecological indicators and monitoring methods for a national monitoring program for ecosystems and species

In the United Kingdom, population trends for a suite of farmland bird species have been established as one of the national government's sustainability indicators (Grice *et al.* 2004). The relevant Ministry furthermore committed to reversing the decline of farmland birds by 2020 as one of their 2000 Public Service Agreement's targets (Grice *et al.* 2004), thus linking explicitly conservation targets to government performance targets.

A significant amount of work has already been done to establish a national framework for NRM targets and monitoring standards in Australia (e.g. Natural Resource Management Ministerial Council 2003; Australian Government 2009). What is still needed, however, is the development of nationally consistent sets of ecological indicators and methodologies to underpin the national frameworks and allow for effective accounting of NRM activities (Possingham 2001; Wentworth Group of Concerned Scientists 2008). The indicators and methodologies need to be quantifiable and repeatable at the scale of site, region, state/territory and nation. As noted by others (Possingham 2001; Wentworth Group of Concerned Scientists 2008), these environmental monitoring and evaluation systems furthermore need to be built on a business model where it is clear to the investors how their funding is being used and whether the conservation 'business' is succeeding or not.

Collaborative development of such ecosystem-based indicators and methodologies (e.g. Ladson *et al.* 1999) would be a significant contribution to community-based citizen science and conservation programs in Australia (e.g. Ryan 2003).

9. Establish an ecological monitoring research program to support Natural Resource Management (NRM) community projects

Most NRM funding bodies stipulate that a component of the project's funds be used for the development and implementation of a MERI plan. Currently, it is the responsibility of each project proponent to develop and implement these plans, usually with some guidance from the funding body. We suggest that it would be more effective if funding bodies apportioned a certain proportion of the project's funds to MERI activities and aggregated those funds into a larger monitoring fund. The accumulated funds would be used to establish a new research program that developed and implemented larger-scale, adaptive-management programs incorporating multiple NRM projects. Using this approach, management interventions and monitoring programs for NRM projects could be established within an experimental framework at the start of the project cycle, and multiple projects could help contribute to a robust research and monitoring framework. Depending on the type of information required, the community group might still be responsible for data collection but the monitoring would be done in accordance with the research group's recommendations. Under this model, projects would not proceed until the MERI plan had been developed by research partners and guidelines for data collection established. We would also advocate that such a research program is funded for a minimum of five years to improve the probability of detecting ecological responses to management interventions.

Conclusion

There is general agreement among researchers, policy-makers and land managers that our environmental programs need to be more accountable (Gibbons *et al.* 2008; Wentworth Group of Concerned Scientists 2008). To date however, the institutional translation of that policy shift into practice has placed a greater administrative burden on the thousands of community groups and individuals implementing NRM programs on the ground, but without necessarily improving conservation outcomes. We propose that a more useful model for a future NRM monitoring framework in Australia should be based around the concept and application of citizen science, whereby institutions provide the necessary research, administrative and educational support to community groups, to enable them to undertake their conservation work in a less burdensome, more valuable manner. In this way, we suggest, there is a greater probability that the longevity of both the monitoring program and community participation will be maintained.

Acknowledgements

Many of the ideas presented here result from a monitoring project undertaken by Trust for Nature in 2007 with funding through WWF and the Australian Government's Threatened Species Network program. Trust for Nature gratefully acknowledges that funding support.

Biographies

Doug Robinson works for Trust for Nature, a statutory conservation organisation, as its Conservation Science Coordinator. For the past 15 years he has primarily worked in an extension capacity, encouraging landholders to implement conservation actions on private land throughout northern Victoria. The monitoring lessons documented here have resulted both from his role with Trust for Nature and a previous role with Birds Australia as the Statewide Coordinator for the Grey-crowned Babbler conservation project.

Lisa Smallbone worked as Trust for Nature's project officer for the Longwood Plains biodiversity monitoring project. She has worked as a research officer over the last five years on a number of landscape ecology projects for the Institute of Land Water and Society, Charles Sturt University. She is currently completing her PhD at Charles Sturt University, investigating bird responses to regenerating agricultural landscapes.

James O'Connor is Birds Australia's Research Manager, and oversees a portfolio of research and conservation projects, many of which involve the mobilisation of large numbers of skilled volunteers. Birds Australia is a 110-year-old non-government organisation dedicated to the conservation of Australia's birds and their habitats.

References

Australian Government (2009). 'Natural Resource Management Monitoring, Evaluation, Reporting and Improvement Framework'. Commonwealth of Australia, Canberra.

Barrett GW, Silcocks AF, Barry S, Cunningham RB and Poulter R (2003). *The New Atlas of Australian Birds*. Birds Australia, Melbourne.

Bradbury RB, Browne SJ, Stevens DK and Aebischer NJ (2004). Five-year evaluation of the impact of the Arable Stewardship Pilot Scheme on birds. *Ibis* **146** (Suppl. 2), 171–180.

Conservation Measures Partnership (2007). 'Open Standards for the Practice of Conservation'. Version 2.0. Conservation Measures Partnership.

Curtis A, Britton A and Sobels J (1999). Landcare networks as local organisations contributing to state-sponsored community participation. *Journal of Environmental Planning and Management* **42**, 5–21.

Fitzsimons J and Wescott G (2001). The role and contribution of private land in Victoria to biodiversity conservation and the protected area system. *Australian Journal of Environmental Management* **8**, 142–157.

Gibbons P, Zammit C, Youngentob K, Possingham HP, Lindenmayer DB, Bekessy S, Burgman M, Colyvan M, Considine M, Felton A, Hobbs RJ, Hurley K, McAlpine C, McCarthy MA, Moore J, Robinson D, Salt D and Wintle B (2008). Some practical suggestions for improving engagement between researchers and policy-makers in natural resource management. *Ecological Management & Restoration* **9**, 182–186.

Grice P, Evans A, Osmond J and Brand-Hardy R (2004). Science into policy: the role of research in the development of a recovery plan for farmland birds in England. *Ibis* **146** (Suppl. 2), 239–249.

Ladson AR, White LJ, Doolan JA, Finlayson BL, Hart BT, Lake PS and Tilleard JW (1999). Development and testing of an Index of Stream Condition for waterway management in Australia. *Freshwater Biology* **41**, 453–468.

Lunt ID, Coates F and Spooner P (2005). Grassland indicator species predict flowering of endangered Gaping Leek-orchid (*Prasophyllum correctum* D.L. Jones). *Ecological Management & Restoration* **6**, 69–71.

Murray–Darling Basin Commission (MDBC) (2006). 'The Barmah–Millewa Forest Icon Site Environmental Management Plan 2006–2007'. Murray–Darling Basin Commission, Canberra.

Natural Resource Management Ministerial Council (2003). 'National Framework for Natural Resource Management Standards and Targets'. <http://www.nrm.gov.au/publications/standards/index.html>

Natural Resource Management Ministerial Council (2010). 'Australia's Biodiversity Conservation Strategy 2010–2030'. Australian Government, Department of Sustainability, Environment, Water, Population and Communities, Canberra.

Olsen P (ed.) (2008). *The State of Australia's Birds 2008*. Birds Australia, Melbourne.

Possingham HP (2001). 'The business of biodiversity'. Tela Report to Australian Conservation Foundation. Australian Conservation Foundation, Melbourne.

Roughley A (2009). 'Developing and using program logic in natural resource management users guide'. Australian Government, Canberra.

Ryan JF (2003). 'Community monitoring – useable data through planning. A review of the processes of implementing data confidence measures for community water quality monitoring'. Papers from Sixth International River Symposium Conference, September 2003, Brisbane.

Watson I and Novelly P (2004). Making the biodiversity monitoring system sustainable: design issues for large-scale monitoring systems. *Australian Journal of Ecology* **29**, 16–30.

Wentworth Group of Concerned Scientists (2008). 'Accounting for nature. A model for building the National Environmental Accounts of Australia'. Wentworth Group of Concerned Scientists, Sydney.

Weston M, Fendley M, Jewell R, Satchell M and Tzaros C (2003). Volunteers in bird conservation; insights from the Australian Threatened Bird Network. *Ecological Management & Restoration* **4**, 205–211.

Wilson CW, Robinson D, van der Ree R, McCarthy M, Vesk P and Saywell S (2009). 'The effectiveness of habitat works for the survival and population status of the Grey-crowned Babbler *Pomatostomus temporalis*'. A report to The Norman Wettenhall Foundation and the Goulburn Broken Catchment Management Authority. University of Melbourne, Melbourne.

PROGRAMS AND THE LESSONS LEARNED FROM THEM

17 BIODIVERSITY MONITORING IN CANADA'S YUKON: THE COMMUNITY ECOLOGICAL MONITORING PROGRAM

Charles J. Krebs

Lesson # 1. Construct a food web for the system under study.

Lesson # 2. You cannot do everything so decide what is important.

Lesson # 3. You can add items to the monitoring list as you go.

Lesson # 4. You must standardise all the measurement protocols and publish them in a small handbook with photos and details of data entry.

Lesson # 5. Maintaining enthusiasm for the monitoring program is critical.

Lesson # 6. Enter the data immediately in the field if possible.

Lesson # 7. Communicate the results as widely and as much as you can.

Lesson # 8. Do not get discouraged.

Lesson # 9. Continue to take a long-term view of monitoring.

Introduction

In 1973 we began studies on small mammal populations in the boreal forest of the south-western Yukon of Canada. By 1984 we had realised that we needed to study community dynamics rather than single species dynamics, and our studies, while still based on experimental manipulations, broadened to monitor the ecosystem. We are now in year 38 of this monitoring program. The lessons we have learned can be condensed into nine key points.

Lessons

1. Construct a food web for the system under study

This exercise forces you to define the biological and geographical limits of what you are trying to monitor, and once the scale is set, practical decisions can be made about how much can be done. The food web for the Kluane ecosystem is shown in Figure 17.1 (Krebs *et al.* 2001). We see right away that things are missing from the food web. We are not monitoring passerine birds, or insects, or the soil fauna, or most of the herbaceous plants. The arrows arise from natural history data, and emphasise that to monitor any ecosystem you must have a great deal of

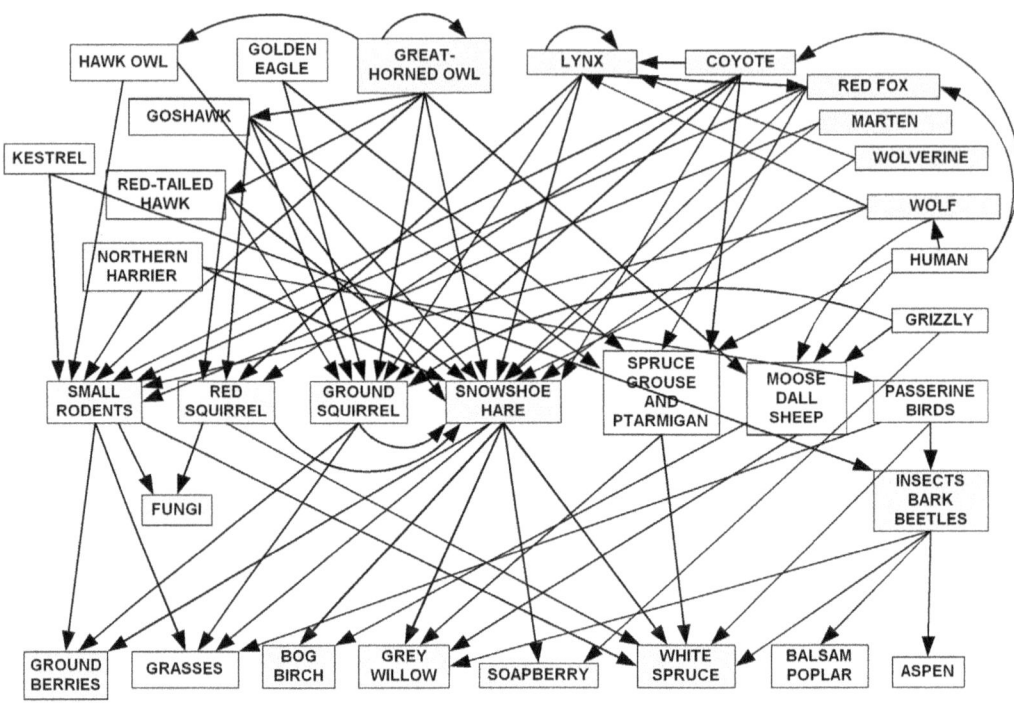

Figure 17.1. The terrestrial food web for the Kluane Lake region of the Yukon, focusing particularly on the mammals. The shaded boxes have been monitored annually for 18–38 years.

background data. Food webs are always constructed from some point of view, and many items are lumped into functional groups or ignored altogether.

2. You cannot do everything so decide what is important

No food web is complete, and all webs can be thought of as partial webs or sub-webs of the entire ecosystem. We have been interested in terrestrial mammal dynamics centred on the snowshoe hare (*Lepus americanus*) cycle, so we have picked elements to monitor in the ecosystem that somehow impinged on hares. We have some understanding of the dynamics of the mammals in the boreal forest, and thus the ability to construct hypotheses that monitoring can evaluate (Krebs 2011). You are also forced at this stage to define the frequency of monitoring and the spatial extent. In our Yukon studies, we range over valley forest sites spanning about 250 km along the Alaska Highway and Haines Road. We do not study the alpine zone; we do not study lakes and streams.

Critical decisions must be made about how many weather variables to monitor and with what level of resolution. If biotic interactions drive the system you are monitoring, less effort may need to be expended to monitor abiotic variables. Conversely, if your system is driven by weather, biotic interactions may be given less attention. These decisions come back to the hypotheses or questions you are addressing with the monitoring program (Lindenmayer and Likens 2010).

3. You can add items to the monitoring list as you go

As we progressed in the work at Kluane we added above-ground mushroom production (Krebs *et al.* 2008), and ground berry production (Krebs *et al.* 2009) to our list of entities being monitored, since these are important food items for small mammals. When the spruce bark beetle began killing old growth white spruce trees, we started monitoring individual trees to measure attack rates with respect to tree age and size. The background to all the measurements in the Yukon is climate change which is rapid and extreme in north-western Canada. The monitoring program must not be set in stone and needs to be reassessed periodically.

4. You must standardise all the measurement protocols and publish them in a small handbook with photos and details of data entry

We started doing this after about 12 years of work once we realised that many different people would be doing the measuring, some with no prior experience, and the principal investigators could not do all the work themselves. The most important item in a monitoring program is to standardise the measurements and to teach new people how to do things in the field. You cannot learn field techniques on a computer. Our monitoring manual and all of our monitoring data are available on my website (http://www.zoology.ubc.ca/~krebs/kluane.html).

We have always adopted the philosophy that our data should be available to anyone for use, as long as acknowledgement is made. Partly because we set up a detailed monitoring handbook, National Parks in the boreal forests of Alaska have taken up our protocols for measuring the same ecosystem components that we measure at Kluane. We are also in the process of expanding the monitoring program to other sites in the southern and central Yukon with the assistance of biologists from the Yukon Territorial Government. We need to monitor whole landscapes rather than single small sites so that we can discuss large-scale patterns of change, and the difficulties of doing this are partly financial and partly having sufficient trained staff who know the protocols.

We have always been suspicious of monitoring programs that are not well standardised with many ecological measurements that are difficult to take precisely. The variance among observers must not be confounded with the variance due to ecological changes.

5. Maintaining enthusiasm for the monitoring program is critical

We found quickly that you could not 'farm out' the hard field work to undergraduates or hired hands, and it is essential for senior investigators and senior postgraduate students to be part of monitoring teams. If you prefer your air-conditioned office to doing field work, your monitoring program will be compromised. Field work allows time to talk to helpers about why we do what we do.

6. Enter the data immediately in the field if possible

We have tried to enter data in the field, if possible on the day they are collected, so that any simple errors can be corrected before they are forgotten. We set up detailed Access™ databases to enter and verify data. This allowed us to do simple and immediate error checks (e.g. this animal's tag number cannot be correct, or this animal was a female when last captured). It is possible to enter data electronically in the field with hand-held devices in some situations, but we are always happy we have a paper copy of data. And always remember to back-up, and then back-up data again.

7. Communicate the results as widely and as much as you can

We do an annual report to update graphs and to discuss time trends in the data. We write popular articles for local newspapers on topics of immediate interest (e.g. very high abundance of rodents this year) and try to explain why some things happen in the local environment. We talk to school groups in the Yukon and the public about our results as much as possible. We have been less successful at public communication than we should have been, partly because with limited funds we prefer to spend the money on field work rather than communication professionals. We publish regularly in the scientific literature and this is essential to maintain the scientific credibility of the work.

8. Do not get discouraged

Funding for monitoring is pathetic and it is easy to get discouraged. It has taken us nearly 20 years to gain the interest of the Yukon Government in supporting and funding a monitoring program. Interesting results for many people only began to appear after about 10 years. Unfortunately, we have not had good rapport with the First Nations people, many of whom are not supportive of our monitoring because they believe that traditional knowledge is sufficient. We have worked hard to change this, and it is slowly improving.

Our federal science funding program in Canada does not recognise monitoring as a valuable research activity so we can continue to work in the field only by not telling them what we are doing. Parks Canada currently shows little interest in our monitoring program and seems to operate on the twin assumptions that Mother Nature will take care of things and that 'if we know grizzly bear numbers' we have an adequate measure of ecosystem integrity. Some government employees feel that counting and measuring mushrooms is not something a 'real man' should do. The stories are endless, but the main point is to persist. The Yukon Territorial Government is currently investing in, and supporting, our monitoring program – a good sign for the future. The public supports monitoring of the environment and in the long run we will be able to show trends that grow more valuable with each passing year.

9. Continue to take a long-term view of monitoring

Ecologists look beyond the next election to see ecosystems on a scale of hundreds of years. Our monitoring programs should have this time frame. This requires an institutional arrangement that does not disappear when the key players retire. I am not sure that this can be done by government agencies, given our present system of short-term governments. The problem remains to be solved. The storage of data for the long term is also a topic of concern to many people. At the moment, I have most of the Kluane Yukon data available in summary form on my website. Where this information will be in 100 years is not known. The problem with ecological data is that much is unreliable unless the methods are clearly defined and statistically rigorous. Consequently, metadata need to be combined with raw data for them to be useful. The rapid turnover of kinds of data storage in the last 20 years gives more worries to the whole issue of data storage. We have raw data from 50 years ago stored currently only because we were able to move from cards to magnetic tapes to 3.5 inch disks to CDs to DVDs before the old technology disappeared.

Conclusions

The Kluane monitoring program is the longest terrestrial ecosystem monitoring program in Canada but it is not secure, and for the present rests too much on a few shoulders. Others in Canada have monitored single species or groups of species, particularly birds and large mammals that can be hunted, and these data sets are valuable. But every analysis of community dynamics has emphasised that much of the action in the ecosystem is in species less charismatic than grizzly bears and moose. What is needed in Australia, as well as in Canada, is an extended discussion of the monitoring problem, what should be monitored, and what the costs will be. Lindenmayer and Likens (2010) begin this discussion. In their book they discuss some of the problems with the large-scale Alberta Biodiversity Monitoring program, which began in 2003, and these critical discussions need to be more common, particularly when programs are being set up. No monitoring program is perfect and we should all seek to improve what we do so we can do it better.

Acknowledgements

I thank all my colleagues from the Kluane Project and all who have worked on the Kluane Ecosystem Monitoring Program over many years. This research has been a collaborative effort, and without the help of Rudy Boonstra, Tony Sinclair, Stan Boutin, the late Jamie Smith, Susan Hannon, Kathy Martin, Roy Turkington, and Mark Dale we would never have achieved our objectives. Our current monitoring is being supported by Alice Kenney, Mark O'Donoghue, Liz Hofer and a host of Yukon students, for whose help I am grateful.

Biography

Charles Krebs is Emeritus Professor of Zoology at the University of British Columbia and Adjunct Professor in the Institute for Applied Ecology, University of Canberra. He has studied the population and community ecology of vertebrates in the boreal forest and tundra ecosystems of northern Canada for 42 years, concentrating on voles, lemmings and snowshoe hares. His scientific passion is to carry out large-scale field experiments to test hypotheses about the ecological processes affecting populations and communities in northern Canada.

References

Krebs CJ (2011). Of lemmings and snowshoe hares: the ecology of northern Canada. *Proceedings of the Royal Society of London, Series B* **278**, 481–489.

Krebs CJ, Boutin S and Boonstra R (Eds) (2001). *Ecosystem Dynamics of the Boreal Forest: the Kluane Project*. Oxford University Press, New York.

Krebs CJ, Boonstra R, Cowcill K and Kenney AJ (2009). Climatic determinants of berry crops in the boreal forest of the southwestern Yukon. *Botany* **87**, 401–408.

Krebs CJ, Carrier P, Boutin S, Boonstra R and Hofer EJ (2008). Mushroom crops in relation to weather in the southwestern Yukon. *Botany* **86**, 1497–1502.

Lindenmayer DB and Likens GE (2010). *Effective Ecological Monitoring*. CSIRO Publishing, Melbourne.

18 MONITORING FOR IMPROVED BIODIVERSITY CONSERVATION IN ARID AUSTRALIA

Chris R. Dickman and Glenda M. Wardle

Lesson #1. Ground-level monitoring allows tracking of species in time and space.

Lesson #2. Remote sensing allows coarse-level tracking of 'big picture' events.

Lesson #3. Monitoring can help to identify key ecological processes and threats to biodiversity.

Lesson #4. Arid Australia has insufficient long-term monitoring sites.

Lesson #5. Ecological thinking should drive decisions on what to monitor.

Lesson #6. Too few people are engaged in monitoring.

Lesson #7. More sites need to be monitored.

Lesson #8. Ground-level monitoring will maximise conservation outcomes.

Lesson #9. Appeal to personal motivations to get more people involved.

Introduction

Achieving cost-effective biodiversity monitoring is a key goal for conservation managers in many parts of the world, but is of particular importance in the arid interior of Australia. On the one hand, arid Australia covers a vast area of some 5×10^6 km^2, comprises diverse mosaics of different habitats and land systems, and supports many endemic species, species-rich assemblages and ecological processes (Morton *et al.* 2011). On the other, arid Australian environments have experienced degradation due to inappropriate agricultural and pastoral practices in many areas as well as the introduction of exotic pests and weeds. These have led to reduced productivity, high rates of species extinction among taxa as diverse as freshwater snails and native mammals, and many more habitats and species that are in decline (Dickman *et al.* 2007). Hence, there is a clear need to establish monitoring programs that will allow us to improve the conservation of biodiversity.

Since 1990, the Desert Ecology Research Group at the University of Sydney has been studying the ecological processes that drive the population and community dynamics of animals and plants in the Simpson Desert in western Queensland. The research began at a single site located in sand dune habitat in the north-eastern part of the desert, but has since expanded to encompass targeted experimental manipulations and to achieve regular or intermittent monitoring at over a dozen sites covering ca. 6000 km^2 from the Northern Territory border to the eastern edge of the desert. Most of the focus is on the soil seed bank, above ground vegetation and vertebrates, although invertebrates also have been subject to considerable study. Rainfall and temperature data are collected by a network of 13 weather stations and wind speed and direction by a single station. Over 200 research papers, books and reports have been produced. The dominant themes of the work have been to uncover the processes that influence and maintain biological diversity, and to try to ensure that our information is used to achieve effective management (e.g. Robin *et al.* 2010). In 2004 and 2006, two of the large properties that had supported our research passed from private hands to ownership by Bush Heritage Australia, a conservation-focused non-government organisation. This now allows the results of monitoring to be translated quickly to conservation management. Radford *et al.* (Chapter 12, this volume) explore this nexus in more detail. Here, we ask what has been learnt in our two decades in the desert, and how monitoring to obtain improved biodiversity conservation might be achieved in the future. Our first three lessons describe key successes in monitoring, the next three outline major deficiencies, and the last three suggest practical solutions.

Lessons

1. Ground-level monitoring allows tracking of species in time and space

Arid environments are renowned for providing 'boom' and 'bust' conditions for biota. Ground-level monitoring of these 'grand natural experiments' provides insight into which taxa respond and the varying magnitudes and mechanisms of response that they show. To cope with the often harsh conditions of the inland, many desert-dwellers persist for long periods in a state of dormancy (e.g. seeds), prolonged metabolic arrest (e.g. freshwater crabs, frogs) or as desiccation-resistant life history stages (e.g. eggs of many invertebrates). Other desert-dwellers are highly mobile and use local or regional areas only during ephemeral periods of plenty (e.g. waterbirds, kangaroos). One-off surveys that are carried out in arid environments during 'bust' periods will fail to sample these organisms: they will be inactive or absent at the time of survey (Figure 18.1). They will also be unable to find the refugia that support the dormant taxa during these periods. Long-term monitoring is therefore essential to confirm their presence. In turn, knowing that species persist is important for interpreting patterns of temporal and spatial change in species composition, building models of their distribution and dynamics, and for understanding adaptations to desert environments more generally.

As one example of the value of ground-level monitoring, we recorded the hairy-footed dunnart (*Sminthopsis hirtipes*) in the Simpson Desert in mid-1992 after 2½ years of intensive live-trapping. This was the first record of this marsupial in Queensland. The 18 months prior to this discovery had been unusually wet, and subsequent bioclimatic modelling suggested that the species' range had expanded eastwards in response to the wetter conditions (Dickman *et al.* 1993). This species, as well as other small marsupials (*Planigale gilesi*, *P. tenuirostris*), rodents (*Leggadina forresti*, *Rattus villosissimus*) and lizards (e.g. *Ctenotus brooksi*, *C. lateralis*, *C. schomburgkii*, *Strophurus elderi*), have appeared and disappeared from our trapping records over 21 years, allowing us to interrogate the factors that affect their populations.

2. Remote sensing allows coarse-level tracking of 'big picture' events

Much of arid Australia is remote and inaccessible, making it difficult to build an extensive network of ground-level monitoring sites. However, with the advent of Landsat satellites in 1972, it became possible to monitor changes in vegetation cover, the extent of flooding, bush-fires and other large-scale phenomena over regional and continental areas. More recent

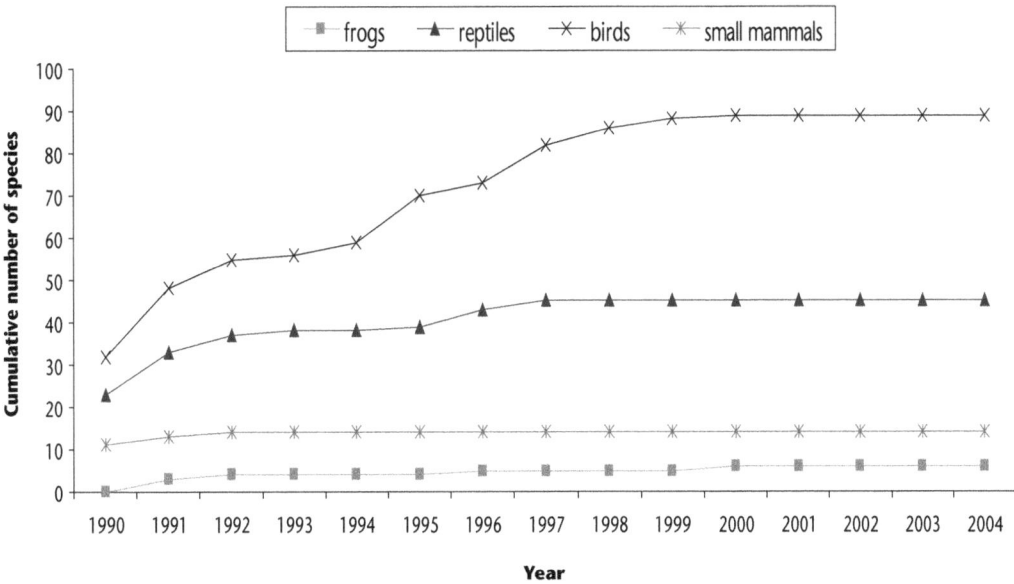

Figure 18.1. Cumulative number of vertebrate species recorded over time at a site in sand dune habitat in the Simpson Desert, western Queensland. The full complement of species in each class took 2½–10 years to be detected, with some of the largest increases taking place after heavy rainfall events. Targeted monitoring took place 4–6 times a year.

satellite-borne sensors achieve very high spatial resolution (1–5 m) and also operate over a broad range of spectral bands that can detect changes in vegetation attributes such as primary productivity and species richness (Kark *et al.* 2008). Of course, remote sensing cannot track the spatial and temporal dynamics of individual species or assemblages, but its ability to monitor the unfolding of 'big picture' events is unparalleled. In the Simpson Desert, Landsat imagery has been used to construct regional-level fire histories and to quantify the propensity of different vegetation types to burn (Greenville *et al.* 2009).

3. Monitoring can help to identify key ecological processes and threats to biodiversity

While tracking biota at different scales is an important goal of monitoring, so too is identifying the biotic or abiotic processes that drive the changes that are observed. When these drivers are understood, managers should be better placed to intervene when necessary to mitigate threats and achieve conservation goals. In our desert work, simultaneous monitoring of biota and weather has confirmed that populations of many species increase shortly after heavy summer rains (>200 mm) but not after light rains (<100 mm). While this might be obvious, the longer-term (1–2 year) consequences of large rains have been anything but expected. Firstly, large rains facilitate the spread of pest and weed species such as cane toads (*Bufo marinus*) and buffel grass (*Cenchrus ciliaris*). They also trigger eruptions of native rodents that in turn support rapid increases in populations of introduced carnivores – especially the red fox (*Vulpes vulpes*) and feral cat (*Felis catus*) – that can then exert heavy predation pressure on populations of native prey species. Finally, large rains stimulate growth of spinifex (*Triodia basedowii*) hummocks and the germination of a diversity of annual grasses and herbs. In the year or so after rain, when this new growth has dried, there may be sufficient ground fuel to support broad-scale and destructive bushfires. Hence, while heavy rains are often seen as the herald of much needed 'boom' times in arid regions, monitoring indicates that they are associated with a range of threats that can place components of biodiversity at risk (Letnic and Dickman 2010).

4. Arid Australia has insufficient long-term monitoring sites

Australia's arid regions are characterised by high levels of biological diversity and almost certainly by different suites of processes that interact to influence component populations and species in different places. But, for the most part, we are uncertain of how well these components of diversity are faring, or what we should do for any that are found to be in decline. There are simply too few sites where consistent monitoring is being carried out. Apart from the ongoing work in the Simpson Desert, dedicated programs of monitoring and experimental research have been undertaken since the mid-1980s as part of the Arid Recovery project in South Australia; monitoring also takes place in a limited number of regional areas to document the effects on native fauna of poison baiting for foxes (most notably in Western Australia; Armstrong 2004), and has been initiated in the last decade on desert properties owned and managed by non-government organisations such as Bush Heritage Australia and the Australian Wildlife Conservancy. There are also a few projects on focal species, often maintained by dedicated individuals (e.g. McRae 2004). The paucity of monitoring sites in arid Australia stands in marked contrast to the situation in parts of Africa and the US, where monitoring for long-term ecological research has been a priority for several decades.

5. Ecological thinking should drive decisions on what to monitor

Because of the insufficiencies noted in lesson #4, and the separate efforts across different funding agencies and state governments, there is no clear consensus about what we should be monitoring in most parts of arid Australia. On the one hand, we might target species or habitats that are known to be threatened. This is being achieved for some high profile species such as waddy wood (*Acacia peuce*) on the edges of the Simpson Desert and the bilby (*Macrotis lagotis*) in the Channel Country, but many of the lesser-known and inconspicuous species remain overlooked. We might also attempt to identify species that have keystone, umbrella, engineering or other important ecological effects, and use them as indicators of the health of the systems to which they belong. Ants, spiders and insectivorous vertebrates have been suggested variously as possible animal indicators, but their functional importance and utility as indicators is likely to vary greatly between desert habitats. Conversely, widespread vegetation alliances such as those dominated by *Triodia* spp. could be monitored remotely to track changes in cover, but this would do little to follow trends in status of the many species that spinifex supports. Perhaps, then, known threats such as weeds, introduced predators and other exotic species should be monitored. This could be done, but on their own would provide little insight into patterns of change in the native biota. On the other hand, we might simply monitor the effects of different management practices or of abiotic drivers of biodiversity change. On-site rainfall, floods, storms, fire and the effects of such events on soil movement and primary productivity could be measured, much of it remotely (e.g. Greenville *et al.* 2009). The effects of livestock removal or of the culling of feral camels (*Camelus dromedarius*) could also be measured using remote means. But such coarse-scale monitoring would inevitably miss the local ground-level changes in species and local assemblages that we wish to conserve. We need to think clearly and explicitly use ecological principles to ensure that our monitoring efforts achieve biodiversity conservation goals.

6. Too few people are engaged in monitoring

The population of arid Australia is relatively small and highly dispersed among many tiny communities and towns. Most people work in the production, mining or service industries, and many move from place to place as different opportunities present themselves (Robin *et al.* 2010). With the exception of environmental managers working on conservation-focused properties such as those of Bush Heritage Australia, the Australian Wildlife Conservancy and Arid Recovery, this often means that monitoring is carried out by state government employees, academics and students who monitor their sites when they can make the often-long journey inland. The Desert Ecology Research Group, for example, needs to allocate 5–6 days on the road for each round trip that members make from Sydney to the Simpson Desert. Despite the small local population base in arid Australia, we have arguably not done enough to support and encourage local capacity. Baker and Mutitjulu Community (1992) made a strong case for combining scientific and traditional knowledge in monitoring via the involvement of Aboriginal people. The results of their work at Uluru National Park suggest that such engagements are highly beneficial, but the model proposed has not been adopted widely. It will be difficult to expand a monitoring program of any kind in arid Australia unless there is a much greater engagement of trained and qualified people.

7. More sites need to be monitored

Getting more monitoring sites established will require more funding from government and private sector sources than is currently available. Although it is challenging to gain extra resources for an enterprise that returns no immediate benefit, it will not happen at all without a vision. We need to specify how many monitoring sites are needed, what they will do, how long they will run, and what benefits they will provide. To do this, we need to set up a decision framework. For example, to determine how many monitoring sites are needed, we could ask:

- Should monitoring sites be stratified by vegetation formation or land system?
- Is there a degree of perceived threat to biodiversity, or some other criterion?
- Should each stratum get one or more monitoring sites?
- Are sites readily accessible at all times?
- Is there human capacity to regularly service the sites?
- Can existing monitoring sites be incorporated into an expanded framework?

The guiding principle for each answer would be whether it meets or exceeds the minimum requirement for improving biodiversity conservation. Answers to questions that are derived using such a decision approach could form the basis of an informed vision that will be essential to leverage improved funding. From a personal perspective, we think that the 12 vegetation formations/land systems with differing rainfall seasonality, as defined by Morton *et al.* (2011), would represent the bare minimum number of site-types that should be established, and would be surprised if any decision analysis were to indicate anything less.

8. Ground-level monitoring will maximise conservation outcomes

Despite uncertainty about what should be monitored, we offer three practical ways forward. First, if we wish to conserve biodiversity and confirm that we are doing so, a program of ground-level monitoring must be implemented at all the sites established in lesson #7 (see also Chapter 14, this volume). A cost-effective general methodology for monitoring a wide range of biota in different arid environments was developed by the Biological Surveys Committee (BSC) (1984) and used successfully over large areas of the interior of Western Australia. This methodology could be adopted more broadly, perhaps with the addition of camera traps. Provided that battery power and memory are sufficient, remotely triggered cameras can be left in place for 3–6 months and often detect rare, cryptic and highly mobile species more cost-effectively than other methods (Towerton *et al.* 2008). They also enable cheaper and more immediate calculation of occupancy and detectability than other methods, and hence 'value-add' to monitoring results.

Another proposal has been made by the Terrestrial Ecosystem Research Network to establish a standard methodology for monitoring the AusPlots-Rangelands network of monitoring sites (Foulkes *et al.* in press). If accepted, this will provide a standardised and repeatable methodology. Second, the monitoring of threatened species should continue in local sites, at least until the processes that impact upon them have been identified and effectively controlled. Third, a wide range of effective remote sensing options is now available, with data often cheaply or freely available (Kark *et al.* 2008), and should be exploited to help achieve conservation goals. Hence, the major investment in what to monitor should be focused on ground-level monitoring, with moderate resourcing allocated to interpreting remotely sensed information.

9. Appeal to personal motivations to get more people involved

Biodiversity motivates people for myriad reasons ranging from economic, personal to social and cultural. Getting people involved in monitoring for better conservation outcomes can appeal to any of these motivations. Economic arguments for conservation monitoring are most persuasive where livelihoods depend on functional systems that can be sustainably exploited. For example, kangaroo populations are routinely monitored to establish an acceptable level of commercial culling based on 15 per cent or less of the estimated population (Grigg 2002). Engaging landholders in gathering information as part of their regular activities is obviously efficient, and even more so if regular summaries were provided to feed directly into adaptive management practices. If pastoralists were required to report annually on stocking rates per paddock, similar to fisheries reporting, that information would accumulate into an impressive resource to inform total grazing pressure and mitigate increased threats to biodiversity through over-grazing in sensitive areas or at critical times (Frank 2010).

Personal motivations for long-term monitoring are evident in existing programs that persist despite short funding cycles (see Lesson #6 above). The strategic involvement of around 600 volunteers has expanded the capacity of the Desert Ecology Research Group to undertake monitoring and has contributed to the training of the next generation of committed researchers. Increasing appreciation of the biodiversity values of remote areas has inspired survey expeditions from groups such as Birds Australia, Australian Geographic, the Royal Geographical Society of Queensland, and sponsored volunteers in Earthwatch programs. An interesting development in arid Australia is the use of camel treks by Australian Desert Expeditions in a style reminiscent of the days of Madigan and the early explorers that crossed the Simpson Desert. Social networks of dedicated enthusiasts already contribute to annual bird surveys, reptile records and plant collections at known locations; with sufficient co-ordination these data archives could be useful for increasing coverage beyond the funded activities. Finally, getting people involved in the future may benefit from new technologies that enable people to report on local observations through web sites and to participate in data refining by screening online databases of images from remotely activated cameras for sightings of target species.

Conclusions

Monitoring for conservation outcomes is best informed by ecological thinking that accounts for more than the simple presence of a limited number of target species. Long-term persistence of communities, and their constituent species, depends on retaining functional processes, consideration of landscape heterogeneity and embracing the dynamic and varying distributions of species. Site-specific data on key environmental drivers such as fire, grazing pressure, feral pests and weeds are critical for interpreting the cause-and-effect relationships for any species or ecological communities of conservation concern.

In our experience from arid Australia, it is not necessary to have annual census intervals to capture the important changes; rather the period of census should be matched to events that stimulate those changes, such as flooding rains or bushfires. This maximises the effective use of limited resources while ensuring that outcomes from interventions or altered threats are quantified and related to identified conservation goals. The most difficult challenge is to shift our perspective on what truly constitutes long-term monitoring. Several years, or even a couple of decades, of data will not be sufficient in highly unpredictable systems. Therefore, ultimate success of all monitoring for conservation outcomes relies on the inter-generational transfer of commitment, non-partisan funding sources, consistent methodologies and excellent systems for data archiving and retrieval.

Acknowledgements

We are indebted to Bobby Tamayo and Aaron Greenville for their major contributions over many years, to large numbers of volunteers for their assistance, to Bush Heritage Australia for facilitating work on their properties, and the Australian Research Council for funding.

Biographies

Chris Dickman is a Professor in Terrestrial Ecology and Australian Research Council Professorial Fellow at the University of Sydney. He has long been intrigued by local and larger-scale patterns of biodiversity and the processes that create them, but remains concerned at the rapidly accelerating rates of biodiversity loss. For the last 20 years his primary focus has been to elucidate, by observation and field experiment, the factors that regulate the diversity of vertebrates and other biota in arid Australia. He is the author of more than 300 papers, book chapters and books.

Glenda Wardle is an Associate Professor in Ecology at the University of Sydney. As a Principal Researcher in the Desert Ecology Research Group she focuses on plants to explore the interactions and processes that underpin the diverse and dynamic desert communities. In her spare time she is partial to number crunching and population modelling.

References

Armstrong R (2004). Baiting operations: *Western Shield* review – February 2003. *Conservation Science Western Australia* 5, 31–50.

Baker LM and Mutitjulu Community (1992). Comparing two views of the landscape: Aboriginal traditional ecological knowledge and modern scientific knowledge. *Rangeland Journal* 14, 174–189.

Biological Surveys Committee (1984). Introduction and methods. In *The Biological Survey of the Eastern Goldfields of Western Australia*. (Ed. KR Newby) pp. 1–19. Western Australian Museum, Perth.

Dickman CR, Downey FJ and Predavec M (1993). The hairy-footed dunnart *Sminthopsis hirtipes* (Marsupialia: Dasyuridae) in Queensland. *Australian Mammalogy* 16, 69–72.

Dickman CR, Lunney D and Burgin S (Eds) (2007). *Animals of Arid Australia: Out on their Own?* Royal Zoological Society of New South Wales, Sydney.

Foulkes JN, White IA and Lowe AJ (in press). AusPlots-Rangelands monitoring site stratification and survey methods within TERN (Terrestrial Ecosystem Research Network).

Frank ASK (2010). The ecological impacts of cattle grazing within spinifex grasslands and gidgee woodlands in the Simpson Desert, central Australia. PhD Thesis, University of Sydney, Sydney.

Greenville AC, Dickman CR, Wardle GM and Letnic M (2009). The fire history of an arid grassland: the influence of antecedent rainfall and ENSO. *International Journal of Wildland Fire* **18**, 631–639.

Grigg G (2002). Conservation benefit from harvesting kangaroos: Status report at the start of a new millennium – a paper to stimulate discussion and research. In *A Zoological Revolution: Using Native Fauna to Assist in its own Survival*. (Eds D Lunney and C Dickman) pp. 53–76. Royal Zoological Society of New South Wales, Sydney.

Kark S, Levin N and Phinn S (2008). Global environmental priorities: making sense of remote sensing. *Trends in Ecology and Evolution* **23**, 181–182.

Letnic M and Dickman CR (2010). Resource pulses and mammalian dynamics: conceptual models for hummock grasslands and other Australian desert habitats. *Biological Reviews* **85**, 501–521.

McRae PD (2004). Aspects of the ecology of the greater bilby, *Macrotis lagotis*, in Queensland. MSc Thesis, University of Sydney, Sydney.

Morton SR, Stafford Smith DM, Dickman CR, Dunkerley DL, Friedel MH, McAllister RRJ, Reid JRW, Roshier DA, Smith MA, Walsh FJ, Wardle GM, Watson IW and Westoby M (2011). A fresh framework for the ecology of arid Australia. *Journal of Arid Environments* **75**, 313–329.

Robin L, Dickman C and Martin M (Eds) (2010). *Desert Channels: the Impulse to Conserve*. CSIRO Publishing, Melbourne.

Towerton AL, Penman TD, Blake ME, Deane AT, Kavanagh RP and Dickman CR (2008). The potential for remote cameras to monitor visitation by birds and predators at malleefowl mounds. *Ecological Management & Restoration* **9**, 66–69.

19 EXPLOITING THE BACK-LOOP OF THE ADAPTIVE CYCLE: LESSONS FROM THE BLACK SATURDAY FIRES

Philip Gibbons

Lesson #1. Successful monitoring does not necessarily have to be underpinned by an *a priori* hypothesis.

Lesson #2. Despite Lesson #1, there must be strong motivations to collect data and maintain data sets.

Lesson #3. Equal emphasis must be placed on monitoring management inputs and biodiversity outcomes.

Lesson #4. Expenditure should be monitored too.

Lesson #5. Monitoring data held by public agencies cannot be accessed quickly enough.

Lesson #6. We need to demonstrate if biodiversity monitoring is a good investment of public money.

Lesson #7. National standards for data collection, documentation and storage are needed.

Lesson #8. There should be a pre-planned response to events that will provide opportunities for biodiversity conservation.

Introduction

The Black Saturday Fires on 7 February 2009 were the worst bushfire disaster in terms of life (173 deaths), infrastructure (2133 houses) and economic cost ($AUD4.3 billion) in Australia's history (Teague *et al.* 2010). Immediately after this event there was a commitment by all parties to learn and do things better and a major inquiry was established to achieve this – a Royal Commission (Teague *et al.* 2010). The immediate period after events of this type represents an important window in which change and innovation occurs (Walker and Salt 2006). This is akin to the 'back loop' of the adaptive cycle (Gunderson and Holling 2002), which is the period in which there is a reorganisation of the system (see Figure 19.1). In this chapter I argue that the quality of data available during this period, the ability to access these data and to analyse them in a timely fashion, and thus turn them into information, are all critical for adaptive learning. I draw on my experience with the Black Saturday Fires to highlight eight observations relating to monitoring in the context of adaptive learning that are relevant for biodiversity conservation.

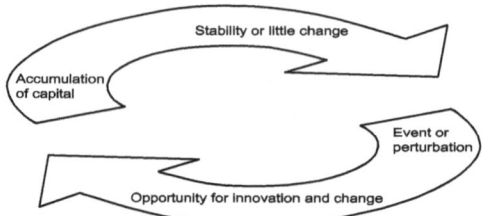

Figure 19.1. The adaptive cycle. Resilience theory predicts that systems cyclically tend towards a 'fore loop' of growth when capital is accumulated (e.g. expertise, funding, biomass, rules) followed by a relatively long period of stability, little change or innovation. External factors (e.g. bushfires, cyclones, political change, financial crises) then precipitate a 'back loop' of upheaval and reorganisation. This back loop represents a potential window – albeit often a small window – of opportunity for data collected during monitoring programs to inform innovation and change (adapted from Walker and Salt 2006).

Lessons

1. Successful monitoring does not necessarily have to be underpinned by an a priori hypothesis

'Omnibus surveillance monitoring' is the collection of longitudinal or repeated-measures data with no *a priori* hypotheses (Nichols and Williams 2006). This type of monitoring has been criticised repeatedly and implicated as a reason for the generally poor longevity of monitoring projects (Lindenmayer and Likens 2010). My experience with the Black Saturday Fires (and elsewhere) has led me to question this critique. Adaptive learning from the Black Saturday Fires was possible only because of the maintenance of a complete spatial and temporal coverage of several data sets collected as part of what can probably be best described as 'surveillance monitoring'. That is, there were no specific questions or hypotheses that underpinned the collection of these data. Nevertheless, these data sets are critical for interrogating important topics, such as bushfires. The availability of these data was critical in the Black Saturday study because events of this type are unpredictable in both their timing and location. What is the alternative to surveillance monitoring in these circumstances? If these data are collected only at sites where there is a current demand for it then opportunities for investigations at different locations are lost. Further, many well-defined monitoring programs draw on, and partly owe their success to, longitudinal surveillance data (e.g. climatic data, satellite imagery). Most long-term ecological studies owe their success to their ability to adapt and change focus, or change the set of hypotheses they address (Lindenmayer and Likens 2009). This suggests to me that motivations other than an *a priori* specific question can underpin good monitoring.

2. Despite Lesson #1, there must be strong motivations to collect data and maintain data sets

Although the data sets I used after the Black Saturday Fires were not collected and maintained with bushfires as the intended application, there were several other motivations for the custodians to collect and maintain these data. First, most of the data sets I used have many different potential applications and their collection is therefore easy to justify – and hard to stop. Second, several data sets were collected because the custodian was obligated or mandated to do so either by legislation or international convention. And third, some of these data sets are available publicly, which immediately makes the custodians more accountable for the maintenance of these data. So, while the existence of these data was not underpinned by specific questions or objectives relating to bushfires, there was at least one strong motivator for the custodians to maintain each of these data sets. This observation leads me to suggest that data for which there are existing and strong motivations on the part of the custodian to collect and maintain should be collected as part of surveillance monitoring programs. An important motivation for other data – data for which no agency is mandated to collect (e.g. data collected as part of research) – is a well-defined objective and specific questions and a conceptual model that is being tested (Lindenmayer and Likens 2010).

3. Equal emphasis must be placed on monitoring management inputs and biodiversity outcomes

Adaptive management requires an understanding of the links between management inputs and outcomes. Management inputs (e.g. grazing regime, burning regime, herbicide application, planting technique) are often the key components of the system that can be manipulated in order to effect a change in outcomes. However, there is greater emphasis on measuring biodiversity outcomes, or finding biodiversity indicators, than suitable ways to measure management inputs, especially among research scientists. Fazey *et al.* (2005) found that only 13 per cent of studies published in journals focused on conservation biology actually tested conservation actions. To some extent this reflects difficulties researchers face collecting good management data. Many forms of management cannot be measured effectively at a single point in time or retrospectively. For example, many disturbances (e.g. grazing, fire) vary with timing, intensity and duration, so require some reasonably detailed information to be gathered over time. Land managers are often in the best position to record these data. The research I undertook after the Black Saturday Fires was only possible because the Government of Victoria collects, and keeps up-to-date, spatially explicit data on several aspects of forest management (e.g. prescribed burning, logging and clearing). All publicly funded land management agencies in Australia should collect, and make available, data on their management actions at a fine spatial resolution and make this a compulsory component for any publicly funded management on private land. Methods to capture these data should be a greater focus of researchers in this field.

4. Expenditure should be monitored too

Establishing links between management actions and biodiversity outcomes is important, but decision-making is also based on the relative cost-effectiveness of management options. In a biodiversity context this will give the marginal gain in biodiversity conserved per dollar spent. Greater biodiversity conservation outcomes can potentially be achieved within a budget if the most cost-effective options for management are identified. This is the thinking that underpins biodiversity auctions or tenders. That is, each proposal to conserve biodiversity is assessed according to its conservation value and each proposal is costed fully. Individual proposals or groups of proposals are ranked on the basis of their cost-effectiveness according to

$$\frac{B}{\$}$$

where B is biodiversity value and $\$$ is cost (Hajkowicz *et al.* 2008). The most cost-effective set of proposals are funded until the budget is expended. Windle and Rolfe (2008) used data of this type to calculate that the same amount of biodiversity could be conserved using one policy instrument (a tender) at 70 per cent of the cost of biodiversity conservation delivered using another instrument (a fixed-price mechanism). It is unacceptable that we cannot compare all biodiversity conservation programs in these terms. There are precious few biodiversity monitoring data sets in Australia in which actual costs of different management options are recorded, so it is not surprising that we have yet to effectively prosecute the argument that monitoring biodiversity is a cost-effective investment!

5. Monitoring data held by public agencies cannot be accessed quickly enough

A key feature of the back loop of the adaptive cycle – or the window of opportunity to institute change after major events – is that this window quickly closes again. A colleague and I took some 12 months to access the data sets we needed to undertake research on the Black Saturday Fires. By the time we had analysed these data the Royal Commission had handed down its final report and the window of opportunity for influencing major change in the way forest fuels are managed had been lost. As part of this research we tested empirically some recommendations of the Bushfires Royal Commission and found they would not meet their objective. Decisions made after major events will be more informed if we can quickly respond to these events with empirical studies. How can this be achieved?

The first change that is required is a shift in the culture of many publicly funded agencies. Many individuals in public agencies take the view that data collected by them are not for use by third parties. This is clearly not the case and is counterproductive to developing new insights. It is unacceptable that the ability to access data from public agencies often depends on the attitude of an individual within those agencies. All publicly funded land management or conservation agencies should maintain a central web-enabled repository of all data sets collected and maintained by them. There is a trend towards this (e.g. Australian Spatial Data Directory http://asdd. ga.gov.au/asdd/, Community Access to Spatial Information (CANRI) http://www.canri.nsw. gov.au/index.html). However, it is still difficult to find a single comprehensive list of data sets held by an agency and how these data can be accessed is not always transparent. Using our experience with the Black Saturday Fires as an example, we had to work through 27 individuals to access land management information for our study area. Finding key individuals took too much time and these people were all under pressure to respond to the Royal Commission, so often did not always place a high priority on our request to access data. An established protocol and dedicated system for disseminating data would avert these problems. We will continue to make ill-informed decisions if we cannot genuinely use the back loop of the adaptive cycle to generate improved knowledge.

6. We need to demonstrate if biodiversity monitoring is a good investment of public money

I believe that a key reason that biodiversity monitoring is not treated with greater priority, does not receive greater funding and is often not implemented effectively, is because we have failed to prosecute the argument that monitoring biodiversity is a worthy public investment. In other words, we have failed to monitor monitoring. The first step in doing this is to arrive at a set of criteria for assessing the effectiveness of monitoring programs. Just because a monitoring program addresses a specific question, is well designed, has produced a long time series of data of high quality and has led to several journal publications is not, in itself, a justification for spending public money to support it. And just because a monitoring program has fizzled out after a short period of time does not, in the broader sense, mean it was a failure. Further, I'm not aware of any monitoring programs that have quantified the cost-effectiveness of the program (but I may be wrong). I believe it is important to assemble a series of case studies from within and without the discipline of biodiversity conservation to illustrate whether improved monitoring is needed for effective biodiversity conservation in Australia.

7. National standards for data collection, documentation and storage are needed

Monitoring programs for biodiversity would benefit from a set of national standards, as has been established in many industries or disciplines (e.g. education, food supply, water, health, climate). The intention of standards is not to make monitoring prescriptive, is not to suggest that one size fits all and is not to inhibit innovation or adaptation. Some key issues that could be the focus of national standards for monitoring, particularly where public expenditure is involved, are listed in Table 19.1.

8. There should be a pre-planned response to events that will provide opportunities for biodiversity conservation

It is inevitable that there will be events (e.g. bushfires, floods, accidents, disease, development proposals, lobbying) that provide windows of opportunity to review, and if necessary change, the way biodiversity is managed. As discussed above, the window of opportunity opened by these events may be small. Therefore, it is important that the data needed to inform such periods of review and innovation should be available to experts with the capacity to derive insights from the information and that these data are accessible in a timely fashion. At present the scientific response to such events is largely ad hoc and uncoordinated. There should be pre-arranged groups of key individuals that can respond to these events at short notice. One example of this is the Burned Area Emergency Response program (BAER) founded in the United States of America that brings together scientific expertise immediately after major fires to undertake a rapid risk assessment relating to environmental hazards. This model has been adopted by Australian bushfire agencies. Similar programs could be organised among scientists and managers in different areas of biodiversity

Table 19.1. Issues that could be considered as part of national standards for biodiversity monitoring. A set of minimum standards for monitoring outcomes should be mandatory where public expenditure is involved.

Issue requiring standard	Explanation
A definition of responsibilities	An understanding of which organisations are responsible for what data would clarify responsibilities and minimise duplication.
Design	Minimum standards around the design of monitoring programs would increase the utility of monitoring data. For example, this standard could specify that independent statistical advice must be sought prior to establishing a monitoring program.
A set of core data	The consistent collection of certain data across organisations and administrative boundaries would improve opportunities for adaptive learning because greater contrasts across environments and management approaches would be sampled. Standards for the types of management data that must be collected by public agencies and the way management costs are recorded are examples where standards could be developed.
Frequency of measurement	Collective knowledge of the rate at which some variables change may dictate a minimum frequency between measurements for certain entities.
Data quality	A standard for data quality might include a minimum acceptable observational error (i.e. error due to differences between observers or recording devices).
Expenditure	For reasons discussed earlier in this chapter, there should be minimum standards for reporting expenditure on biodiversity programs, particularly for actions that are publicly funded.
Metadata	Minimum standards in the way data are documented are highly desirable.
Data accessibility	As discussed earlier in this chapter, this is a key area that requires reform. Public access to all data via the internet should be an aim for public agencies and will improve accountability.

conservation to ensure a timely response to major events in terms of data collection, collation and analysis.

Conclusions

The adaptive cycle predicts that there are periods when there will be windows of opportunity for adaptive learning (Walker and Salt 2006). Unless we have access to high quality data during these often short periods of opportunity and are able to analyse these data in a timely way, then the changes that occur during these periods are unlikely to be based on empirical evidence. In my research after the Black Saturday Fires I was not organised to respond immediately as part of a group of skilled people, the monitoring data needed to do the research took too long to access and some key data (e.g. the cost of different types of fuel management) were not available. Thus, the Royal Commission handed down its findings – and they were accepted by Government – prior to the completion of my research. Interestingly, some of the recommendations made by the Royal Commission were not supported by our results. Decisions made during windows of opportunity that arise after major events will continue to be based on outdated information and rhetoric rather than empirically based evidence unless there is reform in the areas I have highlighted here.

Biography

Philip Gibbons commenced his career in 1988 working with land management agencies in Victoria and then New South Wales on forestry-related management, which included fire-fighting duties. Since his PhD in 1999 he has worked in research positions with CSIRO, the New South Wales National Parks Service and The Australian National University. His research focus is native vegetation management, land-clearing policy and biodiversity conservation and several of his projects rely on sound approaches to monitoring.

References

Fazey I, Fischer J and Lindenmayer DB (2005). What do conservation biologists publish? *Biological Conservation* **124**(1), 63–73.

Gunderson LH and Holling CS (2002). *Panarchy. Understanding Transformations in Human and Natural Systems*. Island Press, Washington, DC.

Hajkowicz S, Higgins A, Miller C and Marinoni O (2008). Targeting conservation payments to achieve multiple outcomes. *Biological Conservation* **141**(9), 2368–2375.

Lindenmayer DB and Likens GE (2010). *Effective Ecological Monitoring*. CSIRO Publishing, Melbourne.

Lindenmayer DB and Likens GE (2009). Adaptive monitoring: a new paradigm for long-term research and monitoring. *Trends in Ecology and Evolution* **24**, 482–486.

Nichols JD and Williams BK (2006). Monitoring for conservation. *Trends in Ecology and Evolution* **21**, 668–673.

Teague B, McLeod R and Pascoe S (2010). '2009 Victorian Bushfires Royal Commission. Final Report'. Parliament of Victoria, Melbourne.

Walker B and Salt D (2006). *Resilience Thinking. Sustaining Ecosystems and People in a Changing World*. Island Press, Washington, DC.

Windle J and Rolfe J (2008). Exploring the efficiencies of using competitive tenders over fixed price grants to protect biodiversity in Australian rangelands. *Land Use Policy* **25**(3), 388–398.

20 WATERBIRD MONITORING IN AUSTRALIA: VALUE, CHALLENGES AND LESSONS LEARNT AFTER MORE THAN 25 YEARS

Richard T. Kingsford and John L. Porter

Lesson #1. Rapid large-scale surveys are achievable, cost-effective and scientifically robust.

Lesson #2. Long-term data sets of waterbird abundance are critical for management of rivers and wetlands.

Lesson #3. Multiple objectives for management and conservation can be met using monitoring data.

Lesson #4. Funding cycles for government do not coincide well with long-term monitoring of populations, a problem exacerbated across multiple jurisdictions.

Lesson #5. Continuity, commitment, responsibility and perseverance are critical to ensure long-term success of surveys.

Lesson #6. Data management and access to long-term data are problematic as there is little investment in ensuring long-term legacy and access.

Lesson #7. Ongoing long-term surveys require training of personnel for data collection as well as succession planning.

Lesson #8. Most government organisations and management agencies have documented commitments to monitoring and evaluation but there is seldom human or financial resourcing provided for collection of these data.

Introduction

Much of the world's freshwater biodiversity is in serious decline (Dudgeon *et al.* 2006; Vörösmarty *et al.* 2010). Evidence of environmental degradation should be the basis for environmental policy and management of all freshwater, terrestrial and marine ecosystems. The more difficult associated question is identifying causes of environmental change, given natural stochasticity of biota and ecosystems. Inevitably, both questions demand long temporal sequences of data, particularly for highly variable environments such as freshwater ecosystems. Generally, abiotic variables for freshwater ecosystems are measured reasonably well at broad spatial and temporal scales (e.g. climate, river flows, water quality), but data for biota are poor, particularly for mobile animals. Waterfowl or wildfowl (Anatidae, ducks, geese and swans) are one group that is routinely monitored around the world, primarily for the management of duck hunting seasons (Kingsford and Porter 2009). This chapter

Figure 20.1. Ten east–west survey bands (solid lines) across eastern Australia used in annual aerial surveys in October of all species of waterbirds over 28 years, 1983–2010. Dashed lines indicate 'one-off' additional survey bands. All survey bands are 30 km wide and every aquatic ecosystem (>1ha) is counted within this band.

describes the value, challenges and lessons learnt for monitoring programs from more than a quarter of a century of aerial surveys of waterbirds, including waterfowl, across about a third of Australia.

Aerial surveys are a preferred method for estimating the distribution and abundance of waterbird populations because large areas can be surveyed at relatively low cost (Bayliss and Yeomans 1990; Kingsford 1999). There have been about 20 major regional or large-scale aerial surveys of waterbird populations in Australia since the 1980s (Kingsford and Porter 2009). The aerial survey of waterbirds in eastern Australia – East Australian Waterbird Survey (EAWS) – is the longest running and most extensive (see Figure 20.1). Its genesis in the early 1980s was in response to the more than 100 000 recreational duck hunting licences issued each year in eastern Australia, with little knowledge of potential effects on game species of waterbirds, mostly duck species. Open seasons for duck hunting in all eastern states and the mobility of waterfowl encouraged collaboration among jurisdictions across eastern Australia (Figure 20.1). In 1983 the EAWS began, and every October (1983–present), all waterbird species (>50 taxa) are counted on wetlands, including rivers, along 10 survey bands (30 km wide), using low-level aerial surveys (Kingsford 1999). At other times, additional survey bands were flown north, south and in between the 10 survey bands (Figure 20.1). During aerial surveys, observers also estimate fullness of wetlands relative to the high water mark, providing an estimate of habitat availability.

Lessons

1. Rapid large-scale surveys are achievable, cost-effective and scientifically robust

Increasingly, there is a need to collect data sets over large spatial scales to inform policy and management, particularly at catchment or jurisdictional scales.

Aerial surveys of waterbirds are rapid and imprecise, but track temporal and spatial change reasonably well; they are also cost-effective compared to ground surveys (Kingsford 1999). Opportunities to collect data for multiple species produce not only data for the suite of water-birds using a wetland, but also allow analysis of functional groups or guilds that can inform on the status and habitat value, or even provide a surrogate for more difficult data to collect on a particular wetland. For example, this could include fish data where the presence and abundance of fish-eating birds on a wetland may be a reasonable surrogate for the abundance of fish populations. Such data can provide valuable information on the ecological status of wetlands and its degree of degradation if combined with other abiotic data analyses.

2. Long-term data sets of waterbird abundance are critical for management of rivers and wetlands

About 70 per cent of Australia is dry with a highly variable climate producing equally variable river flows (Puckridge *et al.* 1998) and wetland inundation (Kingsford *et al.* 1999). Such variability requires long temporal sequences of data to distinguish natural impacts from anthropogenic impacts.

Combined with long-term hydrological data and satellite imagery, the availability of long sequences of waterbird data for particular wetlands has allowed assessment of the impacts of building dams and extracting water on large downstream floodplain wetlands (e.g. Macquarie Marshes, Lowbidgee floodplain). These data are able to track the inevitable dry and wet sequences that produce 'boom' and 'bust' cycles in many wetlands and their dependent biota, but also quantify anthropogenic impacts. Determining impacts of reduced flows to these wetlands can be identified by examining changes to river flows as a result of river regulation and water extraction over long periods of time. It is also possible to identify relationships between river flows and ecological responses over relatively short periods. This allows backcasting of likely impacts on waterbird populations, or other biota, using long-term data sets for river flows. Many rivers have hydrological data extending back before river regulation and extraction, providing opportunities for such approaches to measure anthropogenic impacts on aquatic ecosystems.

3. Multiple objectives for management and conservation can be met using monitoring data

Usually, scientific protocols prescribe the importance of defining a specific question for monitoring but waterbird data have met multiple objectives, not just the original one for which the survey was designed. The broad aim of EAWS in 1983 was to determine the distribution and abundance of game species of waterbirds in relation to duck hunting concentrations occurring in Queensland, New South Wales, Victoria and South Australia but mostly concentrated in the three southern states. Each year, states that currently declare open seasons for duck hunting (Victoria and South Australia) obtain a summary of the survey showing long-term trends, distributions and abundances of hunted species; a breeding index for all species; and distributions of water throughout eastern Australia. This information allows assessments of vulnerability of duck populations to hunting, potential annual recruitment and habitat availability.

Other initially unforeseen objectives have also been met. These include identification of conservation values of wetlands, state of environment reporting, management of non-game species and data informing river policy and management. States and the federal agency have used such information to drive assessments of conservation value of freshwater ecosystems, including gazettal of protected areas (e.g. Paroo overflow lakes, Lowbidgee floodplain, Ramsar nomination). Management for individual species, including determination of vulnerability, have benefited from these data. Finally, long-term declining trends in waterbird numbers, beyond natural flooding and drying cycles, have been instrumental in driving policies to return water to freshwater ecosystems in the Murray–Darling Basin.

4. Funding cycles for government do not coincide well with long-term monitoring of populations, a problem exacerbated across multiple jurisdictions

Most scientific projects are funded for a limited period of 1–3 years, which does not align with a long-term monitoring program. Further, government budget cycles are usually annual and seldom easily align across different states.

The EAWS covers four states (Figure 20.1) and initially was funded as a joint program among the states and the Commonwealth, coordinated by CSIRO. In 1986, the states agreed that it would be coordinated by the NSW conservation agency with five-year funding provided. In the mid-1990s, long-term agreement was established to provide some funding from each of the states of about $5000, with NSW providing all travel costs, data entry costs and salaries for observers. This arrangement continued for more than 20 years, despite rising costs of flying surveys (including doubling of aircraft hire costs). Funding for the surveys was obtained primarily through discussions and agreements by individuals in the various jurisdictions; often these people became champions for the survey. With constant funding difficulties and the timing of the survey being early in the budget cycle, decisions were made to fly the survey without guaranteed funding in some years.

Recently, demands for specifying the true cost of the survey have increased, particularly when management of the survey was transferred to the University of NSW. Funding remains problematic with no agreed position across jurisdictions and the Australian Government. There have been attempts for more than three years to raise the issue to a level of consideration by Environment Ministers but this has yet to be achieved. The most appropriate process would be to establish agreement among the states and the Australian Government at the highest level for five-year fully funded programs. This would allow appropriate planning and investment in data processing systems and appropriate training of personnel.

5. Continuity, commitment, responsibility and perseverance are critical to ensure long-term success of surveys

Long-term surveys require funding, people, planning and – most of all – commitment. If they extend over large spatial scales they can be even more challenging. Ensuring one or a few key individuals are responsible and are supported can help ensure the longevity of long-term surveys. For government agencies, this means funding permanent positions dedicated to monitoring. Such long-term involvement ensures that there is continuing understanding of how data are collected and understanding of the uncertainties in using these data. Further, these individuals can ensure that methodology remains consistent over the survey, allowing for comparisons over time. With changing personnel, there can be the temptation to improve survey methodology but this renders comparison with past data impossible. Methodology for the EAWS has remained the same since its inception, although navigation, data recording and processing methods have improved considerably.

6. Data management and access to long-term data are problematic as there is little investment in ensuring long-term legacy and access

Data management and access to long-term data are problematic. There is usually little investment in ensuring access and ensuring quality assurance exists. Further, licensing issues are seldom adequately addressed.

Data storage of biotic data sets remains a serious problem. There is usually little government commitment to improving databases or ensuring quality assurance of current data. Funding is usually difficult enough for obtaining the data but maintaining databases is usually an even lower priority. The EAWS database is reasonably well maintained but inadequate resources for development and maintenance mean that some data are not sufficiently validated. Not all data have been adequately archived. The main part of the database includes the ten survey bands but in other years additional surveys bands were flown (Figure 20.1; Kingsford and Porter 1999). Databases for the EAWS were a corporate data set within the conservation agency but some data for some of these additional survey bands were lost during an upgrade of hardware and software where priorities were for operational data systems (e.g. financial and human resource management).

There is a need to provide funding to ensure that long-term data sets are stored and managed well. This includes identifying ways in which data may be accessed. Large biotic data sets such as the EAWS need to be supported by governments so that they may be adequately maintained and accessed when required, preferably through a guided interface, allowing inquiries for obtaining data remotely. A subsidiary issue is ensuring a system of licensing so that access is regulated and there is opportunity for researchers who collect these data to be given time for analysis and writing up.

7. Ongoing long-term surveys require training of personnel for data collection as well as succession planning

Inevitably personnel change throughout the duration of long-term surveys. There is a need for permanent skilled personnel as well as temporary skilled personnel for surveys. A key challenge is to ensure new personnel can be trained and there is succession planning, particularly if a survey is planned to be ongoing.

EAWS has more than 20 trained observers, of whom only about four are currently willing and able to do survey work. Others have either moved to other jobs or are not allowed to spend two weeks on aerial survey away from their current responsibilities. Training is an exacting process with more than 50 hours required to ensure that an observer can identify and count waterbirds. A decision often needs to be made on whether a particular individual is likely to be able to do future surveys, although they may have the necessary skills and disposition. There has been little thought given to succession planning, largely because currently there is no funding stability for annual surveys. It will be critical for the longevity of the EAWS data set that it is eventually embraced by, and becomes core business within, government. This is problematic for surveys such as EAWS that extend across jurisdictional boundaries.

8. Most government organisations and management agencies have documented commitments to monitoring and evaluation but there is seldom human or financial resourcing provided for collection of these data

There are clear responsibilities for most management agencies to determine effectiveness of conservation management. This is well recognised, with processes for monitoring and evaluation well established, but seldom are these processes adequately resourced.

All states, territories and the Australian Government have responsibilities, under various legislative instruments and agreements, both for determining the state of ecosystems and managing threats. For example, the Australian Government has to report on the ecological character of Ramsar-listed wetlands (DEWHA 2008). Most conservation policy makers and managers recognise the need to monitor ecosystems and biota, but are daunted by the demands of surveying the numerous species and ecosystems. As a result, monitoring and evaluation processes are put in place but seldom adequately funded. Identification of variables to monitor should reflect the values and priorities of policy makers, managers and the public and be responsive to change over large spatial and temporal scales. Ideally, they should be tied to objectives within an adaptive management context (Kingsford *et al.* 2011).

Conclusions

Abiotic data are routinely collected over large spatial scales but these need to be informed by biotic data. Collection of data for conservation over large spatial scales and long periods is taxing in resources and individual input. This is why there are so few long-term data sets. Yet, there is a pressing need to collect such data if deleterious policies and management are to be removed or when rehabilitation occurs and success can be measured. Currently, significant funding is provided for natural resource management and on-ground works to rehabilitate environments, usually at relatively small scales. The success of such works is seldom measured and there is no assessment of whether financial investments are providing a good return. Further, there is a pressing need for broad and often politically difficult conservation policy decisions in the face of strong industry pressure. Such policy will not be successful without good biological data on trends in biodiversity and other measures of the environment.

There are many challenges that need to be overcome to ensure long-term collection of data. The first assessment that needs to be made is whether a monitoring program is responsive to the threat measured. It is also important, given limited resources, to maximise the spatial scale of data collection, making it clear that it needs to reflect the scale of management. Without long-term biotic data, Australia's considerable investment in environmental rehabilitation and protection policy and management cannot be adequately measured to ensure it is achieving its aims.

Acknowledgements

We thank all those whose dedication to aerial surveys of waterbirds has so far ensured that we continue to collect long-term data. This includes the many observers and pilots who have helped in the collection of aerial survey data across about a third of the continent over a period of 28 years so far.

Biographies

Richard Kingsford is Director of the Australian Wetlands and Rivers Centre at the University of NSW. He spent 18 years in the NSW conservation agency grappling with the practicalities of monitoring, policy and management. His research focuses on the waterbirds, wetlands and rivers of Australia, particularly investigating impacts of water resource development in the Murray–Darling Basin. He advises various governments through committees on water, wetland and river management.

John Porter is a Research Scientist in the Parks & Wildlife Group within the Department of Environment, Climate Change and Water. He has been involved in waterbird and wetland research and monitoring for 22 years. His research interests include waterbirds, aquatic plant communities (including charophyte algae) and measuring management effectiveness.

References

Bayliss P and Yeomans KM (1990). Seasonal distribution and abundance of Magpie Geese *Anseranas semipalmata* Latham, in the Northern Territory, and their relationship to habitat, 1983–86. *Australian Wildlife Research* **17**, 15–38.

Department of the Environment, Water, Heritage and the Arts (DEWHA) (2008). 'National framework and guidance for describing the ecological character of Australia's Ramsar wetlands. Module 2 of the National Guidelines for Ramsar wetlands – Implementing the Ramsar Convention in Australia'. Australian Government Department of the Environment, Water, Heritage and the Arts, Canberra. <http://www.environment.gov.au/water/publications/environmental/wetlands/pubs/module-2-framework.pdf> [Accessed 15/3/11].

Dudgeon D, Arthington AH, Gessner MO, Kawabata ZI, Knowler DJ, Lévêque C, Naiman RJ, Prieur-Richard AH, Soto D, Stiassny MLJ and Sullivan CA (2006). Freshwater biodiversity: importance, threats, status and conservation challenges. *Biological Reviews* **81**, 163–182.

Kingsford RT (1999) Aerial survey of waterbirds on wetlands as a measure of river and floodplain health. *Freshwater Biology* **41**, 425–438.

Kingsford RT, Curtin AL and Porter JL (1999). Water flows on Cooper Creek determine 'boom' and 'bust' periods for waterbirds. *Biological Conservation* **88**, 231–248.

Kingsford RT and Porter JL (2009). Monitoring waterbird populations with aerial surveys – what have we learnt? *Wildlife Research* **36**, 29–40.

Kingsford RT, Biggs HC and Pollard SR (2011). Strategic adaptive management in freshwater protected areas and their rivers. *Biological Conservation* **144**, 1194–1203.

Puckridge JT, Sheldon F, Walker KF and Boulton AJ (1998). Flow variability and the ecology of large rivers. *Marine and Freshwater Research* **49,** 55–72.

Vörösmarty CJ, McIntyre PB, Gessner MO, Dudgeon D, Prusevich A, Green P, Glidden S, Bunn SE, Sullivan CA, Reidy Liermann C and Davies PM (2010). Global threats to human water security and river biodiversity. *Nature* **467**, 555–561.

21 BIODIVERSITY MONITORING IN THE AUSTRALIAN RANGELANDS

I. Andrew White, Jeff N. Foulkes, Ben D. Sparrow and Andrew J. Lowe

Lesson #1. The vastness of the Australian rangelands makes monitoring difficult, particularly in light of limited resources.

Lesson #2. Australian rangelands have huge climatic and spatial variability, making continental monitoring programs challenging to develop, implement and maintain.

Lesson #3. There are huge gaps in data, knowledge and skills in the Australian rangelands.

Lesson #4. Monitoring of the Australian rangelands requires a national focus and should encourage stakeholder involvement through collaboration.

Lesson #5. Rangeland monitoring methods need regular review to ensure ongoing adaptation as new information becomes available.

Lesson #6. To maximise effectiveness, monitoring in the rangelands needs to be event-driven rather than seasonally focused and therefore should be implemented long term.

Lesson #7. Technological developments have greatly advanced the understanding of rangeland systems through improved monitoring methods, data collection, data analysis, data storage and spatial analysis and these developments need to be continually implemented.

Lesson #8. Biodiversity data should be quantitative and collection methods standardised wherever possible, to facilitate data use from a wide variety of sources.

Introduction

The significant characteristics of the Australian rangelands relevant to biodiversity conservation and management are their considerable size (see Figure 21.1), broad diversity of climate, soils and ecosystems, and extreme temporal and spatial variability. Arid rangelands are generally identified as having historically high rates of species decline attributed to a variety of factors (e.g. overgrazing by livestock, feral herbivores and predators, altered fire regimes; Morton 1990; Short and Smith 1994; Read and Bowen 2001; Johnson *et al.* 2007; Kerle *et al.* 2007; McKenzie *et al.* 2007), with this being acknowledged more recently as continuing in the tropical northern regions (e.g. Woinarski *et al.* 2010). The extent of the rangelands, coupled with a limited and sparse population and a paucity of skilled and experienced ecologists, presents issues of scale for undertaking surveys, surveillance and monitoring, resulting in often limited understanding and ineffective management. With the sparse human population

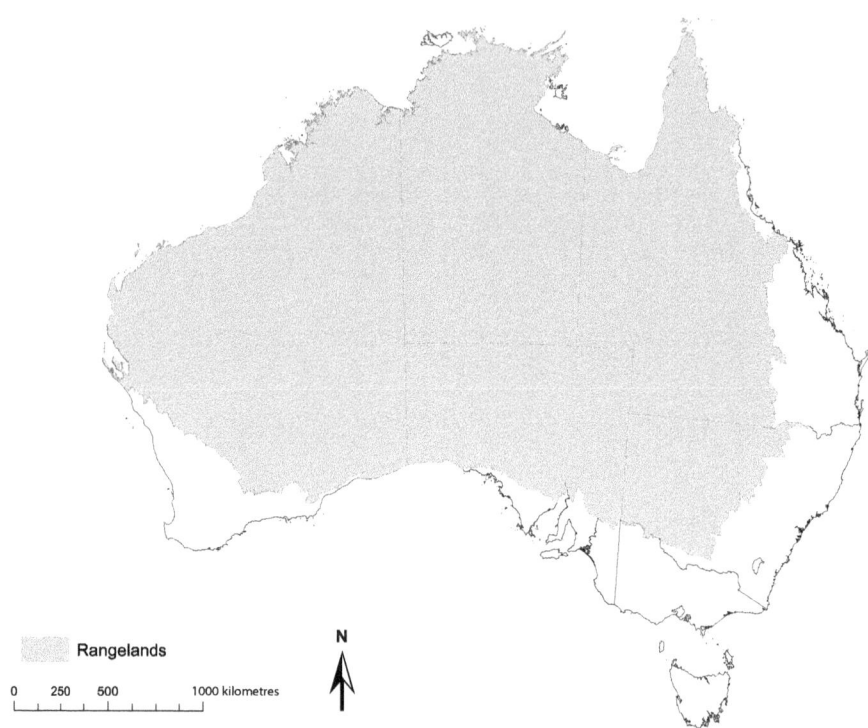

Figure 21.1. The Australian rangelands (after Bastin *et al.* 2008).

and perceived remoteness, there is little political imperative to provide resources to the region. The remoteness of rangelands also means costs of conducting biological monitoring activities are significantly higher than in the more populous temperate and coastal areas.

Historical and contemporary monitoring activities in the Australian rangelands have been extensive, largely targeted at assessing the localised effects of grazing domestic herbivores on native pastures, and as a result are limited in their ability for reporting on broader biodiversity or conservation issues (Bastin *et al.* 2008). There are few biodiversity benchmarks throughout much of the rangelands against which to gauge ecosystem shifts due to changes in management, environment or climate.

In instigating a new biodiversity monitoring program, important lessons have been learned from the extensive monitoring activities already undertaken within the rangelands both in Australia (Whitehead *et al.* 2001; Watson and Novelly 2004; Watson *et al.* 2007a;b) and overseas (Spaeth *et al.* 2003). With the advent of TERN AusPlots we have attempted to incorporate the learning from the past into developing our baseline surveillance monitoring system (see Box 21.1). The objective of AusPlots Rangelands (Foulkes *et al.* 2011) is to establish permanent plots throughout the Australian rangeland bioregions where a baseline assessment of vegetation and soils will be conducted. AusPlots aims to develop and implement a stratification process to decide the location of plots, which is applicable at a continental scale and develop standardised plot assessment methods to be used for measuring and sampling vegetation and soils for a wide range of applications (see Table 21.1). The overall TERN objective is to provide a national institutional infrastructure network for terrestrial ecosystem research and management.

Box 21.1:   TERN AusPlots – an example of learning from past lessons in the rangelands

Overview

The Terrestrial Ecosystem Research Network (TERN) is an initiative of the Australian Government conducted as part of the National Collaborative Research Infrastructure Strategy and the Education Infrastructure Fund Super Science Initiative. The overall objective of TERN is 'to provide a national institutional infrastructure network for terrestrial ecosystem research and management'. TERN is an extensive collaboration with universities, research groups and governments across the Australian biodiversity landscape. Through these different facilities, TERN will act as the overarching and integrative national framework for infrastructure that will facilitate better understanding and management of biodiversity and natural resources in Australia.

AusPlots-Rangelands is an initiative within the Terrestrial Ecosystem Research Network and its objective is to establish permanent biodiversity plots throughout many Australian rangeland bioregions where baseline surveys of vegetation and soils are needed. Initial assessments will incorporate historical monitoring activities, where possible, and will be conducted with readily repeatable methods to enable future monitoring and facilitate temporal comparisons. This will be achieved by developing and implementing a stratification process, applicable at a continental scale, to decide the locations of plots, and developing standardised plot assessment methods to be used for measuring and sampling the vegetation and soils at these locations.

Lesson 1. AusPlots-Rangelands stratifies Australia's rangelands and then selects which bioregions to sample based on their environmental variability to ensure sampling is conducted across the diversity of Australian rangeland bioregions.

Lesson 2. The project stratifies Australia based on climatic variability, landscape and major vegetation groups. Bioregions to sample are then determined by the stratification so that we ensure that samples are representative across Australia.

Lesson 3. The project aims to fill some of the extensive gaps by locating the majority of plots in areas that have not been previously surveyed.

Lesson 4. AusPlots-Rangelands is a project that encompasses all of Australia's rangelands. Survey methods and a stratification process have been developed to encompass the variability throughout the rangelands. These processes are conducted collaboratively with state and federal agencies, private companies, Australian Collaborative Rangelands Information System (ACRIS) and citizens with interest in rangeland areas.

Lesson 5. AusPlots has implemented an adaptive management framework so that we will regularly review our methods and the needs of the stakeholder base and adapt as required.

Lesson 6. Once our plots are established and our baseline data collection completed we will encourage continued monitoring by state agencies, particularly based on significant environmental events, though it is acknowledged that this is outside of the funding and time scale of the current project. It is proposed that further funding be sought for continuation, and to expand the surveys to include faunal biodiversity monitoring.

Lesson 7. AusPlots surveys will make use of the latest technologies by developing, in collaboration with our project partners, digital data collection, next generation photo point collection methodology and analysis procedures, making use of modern plant and soil genetic techniques and collaborating with AusCover to validate Australia-wide ecological products derived from satellite imagery. Data will be stored and disseminated in a modern web based database being developed by the co-located Eco-informatics facility.

Lesson 8. Data collection methods used for AusPlots-Rangelands are quantitative, and the methods have undergone widespread review with the intention of maximising the utility of the data to the widest possible base within the ecological community.

TERN AusPlots has been developed and implemented keeping in mind the valuable lessons learnt by the many skilled ecologists and rangeland scientists that have been and still are working in Australia's rangelands. Although we've attempted to address each of the lessons identified in this chapter we've also adopted an adaptive management framework to regularly review and, where necessary, adapt our methods and approach as new information becomes available.

Lessons

1. The vastness of the Australian rangelands makes monitoring difficult, particularly in light of limited resources

The Australian rangelands are vast, covering around 81 per cent of the continent – more than 6 million km^2 (see Figure 21.1). The sheer size and subsequent variability of ecosystems and climates (from summer dominant monsoonal in the north, through arid in the centre, to winter dominant Mediterranean in the south) has produced significant challenges in assessing and understanding biodiversity and its drivers. Biodiversity monitoring activities must be matched to available financial, temporal and personnel resources as well as logistical constraints. In light of the limited resources, the location of monitoring activities must be targeted to significant areas decided through a transparent stratification process. Monitoring methods need to be consistent but also have a level of flexibility to adapt to different situations.

2. Australian rangelands have huge climatic and spatial variability, making continental monitoring programs challenging to develop, implement and maintain

Seasonal conditions are highly variable across the Australian rangelands, exacerbated in areas south of the influence of the monsoonal tropics. Native plant and animal species are adapted to this variability and are able to survive extended periods of stress caused by drought and/or fire. With the introduction of additional stressors of grazing (by either large numbers of domestic herbivores and/or uncontrolled feral herbivores), and efficient new predators, dramatic biodiversity declines and extinctions have been recorded (Morton 1990). Without monitoring programs it can often take many years for the species declines to be recognised, by which time it can be too late to take action (Wintle *et al.* 2010).

Monitoring of biodiversity in the rangelands needs to be able to separate the effects of seasonal variability from anthropogenic stresses. This is more likely to be achieved through an appropriate experimental design and interpretation of long-term data sets with biodiversity fluctuations related to seasonal conditions and management activities.

3. There are huge gaps in data, knowledge and skills in the Australian rangelands

The provision of permanent biodiversity monitoring plots will increase the likelihood of additional uses of monitoring information. Infrastructure does not need to be extensive, but protection from future anthropogenic disturbances needs to be ensured to enhance the continuing value of plots. Location of the monitoring plots is critical to increase their ongoing use. Scale issues arise, such that monitoring plots should encompass areas that are representative of extensive ecosystems, as well as more restricted components considered significant for catchment function or refuge areas of particular species or guilds of species.

4. Monitoring of the Australian rangelands requires a national focus and should encourage stakeholder involvement through collaboration

Historical monitoring activities have largely used inconsistent methods and field protocols and thus produced information incompatible for spatial comparisons. Previous monitoring has been undertaken by state and territory jurisdictions with little consistency in methods. From a national perspective, a consistent baseline surveillance monitoring program should be implemented to enable regional comparisons.

Collaboration is relevant in both the design of monitoring methods and implementation of specific monitoring activities. By involving a wide range of current and potential participants in the biodiversity monitoring process, the possibility of productive partnerships is enhanced and the likelihood of ongoing adoption increased. By using an inclusive approach from the planning phase through the operational phase to final reporting, it is likely that the different agendas of participants will be addressed and the overall value of the monitoring activities enhanced. This is highlighted in the AusPlots-Rangelands monitoring by the wide-ranging applications (Table 21.1).

5. Rangeland monitoring methods need regular review to ensure ongoing adaptation as new information becomes available

Substantial monitoring efforts have occurred in the different rangelands jurisdictions, targeted at either discovering the impacts on palatable plants from grazing (state/territory-based pastoral monitoring programs) or investigating area specific or species specific issues (Long Term Ecological Research (LTER) and other long-term monitoring activities; e.g. Letnic *et al.* 2005). Grazing monitoring has proven informative in depicting change in vegetation cover, composition and structure from grazing, but generally of little value for biodiversity conservation, while the specific monitoring has provided valuable information for isolated areas and in most instances rare vertebrate taxa. Numerous important lessons have been learnt from these monitoring activities, the most important being that to increase the likely success of a monitoring activity, clear targets must be identified. Other lessons learnt include the required duration of monitoring programs, the flexibility needed in the interval between monitoring events, as well as location of monitoring plots in appropriate places in the landscape (Kerle *et al.* 2007).

An adaptive management approach should be adopted for the methods, with flexibility that includes a constant review process so that methods can be adapted to conditions and recent developments.

Table 21.1 Survey parameters, methods, indicators and applications for AusPlots–Rangelands surveys.

Survey parameter	Parameter element	Measurement method	Indicator	Application
Vegetation: - perennial - annual - ephemeral	- trees - shrubs - forbs - grasses - vegetation cover - substrate cover	identification to species level	- species distribution - species abundance - species/community persistence - site condition - site variability	- population structure - demographic profile - condition assessment - species distribution and dynamics modelling (e.g. bioclimatic envelope predictions)
		point intercept at species level	- site diversity measures - cover of individual species - total cover (with annuals) - species abundance - population vertical structure - species/community persistence	- population structure - demographic profile - condition assessment - remote sensing validation - geographic extrapolation of interpretations - temporal extrapolation of interpretations - retrospective: refine/enhance/ accuracy assessment for AusCover products
		Leaf Area Index meter	Leaf Area Index	- remote sensing validation - retrospective: refine/ enhance/ accuracy assessment for AusCover products
		1 x voucher specimen collected for each species in a plot, with leaf samples removed for genetic analyses	- taxonomic identification and tracking taxonomic change through the data - DNA barcoding - vegetation carbon content	- biodiversity discovery - phylogenetic taxonomy - carbon budget changes
		Photopoint (new method being developed)	- life form - site patchiness - population height/vertical structure - plot basal area	- population structure/ structural complexity - population dynamics - demographic processes - condition assessment (contributing information)

Survey parameter	Parameter element	Measurement method	Indicator	Application
Vegetation: - perennial only	- trees - shrubs - forbs - grasses	5 x leaf specimens collected per plot for dominant species	- DNA barcoding and genomic analysis	- genetic assessment of population connectivity and community diversity
Soils	soil samples	MIR and wet-chemistry per plot : 9 x soil cores to 30 cm, 1 x soil core to 1 m+	- soil nutrition (N, P, K etc.) - soil carbon - bulk density	- enhanced knowledge of rangelands soils - carbon and other nutrients budgets
	soil crusts	soil metagenomics (genetic material from environmental samples)	- taxonomic identification - DNA barcoding for population genetics	- microbial and cryptic soil biodiversity - genetic assessment of population connectivity and community diversity
		cryptogam identification	- taxonomic identification - DNA barcoding for population genetics	- biodiversity discovery - genetic assessment of population connectivity and community diversity

6. To maximise effectiveness, monitoring in the rangelands needs to be event driven rather than seasonally focused and therefore should be implemented long term

Ecosystem processes in the rangelands are largely driven by stochastic climatic events. While regular monitoring activities will provide some data, it is preferred that biodiversity monitoring activities should be event driven, responding to suitable climatic conditions, not to a rigid schedule. To a large extent this will be influenced by location, the areas with more regular seasonal patterns (i.e. the tropics) require less flexibility than the highly variable arid zone. This flexibility conforms to neither accepted funding cycles nor staff management practices and would require a major paradigm shift from organisations that provide support for monitoring activities in the rangelands.

7. Technological developments have greatly advanced the understanding of rangeland systems through improved monitoring methods, data collection, data analysis, data storage and spatial analysis and these developments need to be continually implemented

Technological advances can greatly enhance the effectiveness of biodiversity monitoring, both through the measures provided by the new technology (see Table 21.1) and as a tool to direct more traditional monitoring efforts. Remote sensing as an interpolative tool is a prime example where continental measures such as the recently developed MODIS fractional cover products (Guerschman *et al.* 2009) provide valuable interpretive data, while mapping of fire scars or extent of flooding events can be used to direct on-ground vegetation monitoring activities or aerial surveys for waterbird breeding events.

The collection of leaf samples from plant species and soil crust samples across the rangelands provides an opportunity to undertake chemical and isotope analysis on such material to estimate a range of ecosystem health, carbon storage and nutrient stock indices (e.g. Wynn and Bird 2008). Similarly DNA barcoding and environmental metagenomic analysis of leaf and soil samples can provide important information on phylogenetic diversity, community drivers, genetic refugia, genetic connectivity, cryptic biodiversity and gene expression (e.g. Lozupone and Knight 2007; Graham and Fine 2008; Pavoine *et al.* 2011).

Another exciting area of development is the potential to integrate a series of photographs from a location to derive three-dimensional structural information from a site (this can also be derived from new ground-based LIDAR approaches, e.g. Echidna), or to interrogate temporal or spatial sequences of photographs from a single site to automatically describe vegetation cover, composition and structural changes over time and/or space. Enhanced data collection, management, storage and availability through software development and the creation of web-based portals have streamlined the collection and analysis of biodiversity monitoring data. Ongoing development is occurring through initiatives such the Atlas of Living Australia (http://www.ala.org.au) and the TERN Eco-informatics facility (http://www.tern.org.au).

8. Biodiversity data should be quantitative and collection methods standardised wherever possible, to facilitate data use from a wide variety of sources

Due to the scale issues already identified for the rangelands, the utility and value of collected data need to be maximised. Previous categorical data have proved difficult in interpretation, so wherever possible, data should be quantitative. To facilitate spatial and temporal comparisons data need to be collected using standard readily repeatable methods. For example the Aus-Plots-Rangelands (Foulkes *et al.* 2011) will establish sampling plots that will be 1 ha in area, i.e.

100 x 100 m, with the corners located using DGPS and permanently marked. Each plot will have 10 x 100 m point intercept transects to determine vegetation species, cover and growth form, and 9 point locations. Substrate, plant details i.e. species name, height, cover, strata (as per National Vegetation Information System (NVIS)), flowering, fruiting or senescence and vertical structure will be recorded at 1.0 m intervals along the 10 x 100 m transects. The point intercept method will be used as this is the preferred and most repeatable vegetation cover measurement method for all vegetation strata (Godínez-Alvarez *et al.* 2009; Vittoz and Guisan 2007), which enables the calculation of vegetative cover of each species.

Conclusions

Because of the spatial extent of the Australian rangelands, knowledge of biodiversity will always be a compromise between the desired level of information and the effort and expense in obtaining it. AusPlots-Rangelands will facilitate and encourage the filling of biodiversity knowledge gaps by gathering comprehensive data on soils and vegetation to inform the current status of biodiversity and provide baseline data for future assessments. AusPlots-Rangelands aims to establish up to 1000 plots (1 ha) across the rangelands where baseline measures of soils and vegetation will be taken and samples collected for additional analyses. The information collected will have wide-ranging uses. These will include the more conventional outcomes such as improving the knowledge of the distribution and dynamics of soils and vegetation across the rangelands, as well as for validating remote sensing, improving knowledge of carbon dynamics, and climate change, and providing opportunities for detailed DNA analyses of rangelands plants. Working in collaboration with other initiatives e.g. ACRIS (at http://www.environment.gov.au/land/rangelands/acris/) and Bush Blitz, will value-add to the survey activities producing more complete data sets.

Positive biodiversity outcomes in the rangelands need to be encouraged and adopted where possible and biodiversity champions should be acknowledged for the contributions they make to the understanding and appreciation of Australian ecosystems. These will all enhance the return on conservation investment in the coming 10–20 years. A few examples of these include the Indigenous ranger programs, both aquatic and terrestrial, species recovery projects such as Arid Recovery (http://www.aridrecovery.org.au/) and Operation BounceBack (e.g. de Preu and Pearce 2006), the research and management champions present throughout the region, and positive, innovative organisations such as the Remote Economic Participation (formerly Desert Knowledge) Cooperative Research Centre that highlight the innovative outcomes that emanate from the remote regions of Australia.

Biographies

Andrew White has worked in the Australian rangelands since 1986 on a variety of research, extension and management projects in the Northern Territory (fire, shrub thickening, weeds, feral animals), Queensland (soils and pasture response to floods in the channel country) and South Australia (Executive Officer of the SA Arid Lands NRM Board). He has a BSc from University of Adelaide, a MAppSc from UNSW and is in the final stages of a PhD from UQ investigating grazing management in the tropical savannas of northern Australia.

Jeff Foulkes has worked widely in biodiversity survey, vegetation mapping, ecological research and monitoring in South Australia and the Northern Territory over the last 25 years, much of it in rangelands. Jeff undertook research on the ecology of Common Brushtail Possums in central Australia as well as investigations into the relationships between flora and fauna species and concentrations of soil nutrients and moisture in the central deserts of the Northern Territory. More recently he has coordinated the biodiversity survey and vegetation-mapping program across South Australia, with particular emphasis on the sandy deserts in South Australia.

Ben Sparrow has worked for the past 15 years in many roles for both the South Australian Department of Environment and Natural Resources and the Northern Territory Department for Natural Resources, Environment, the Arts and Sport. He is now survey manager for the TERN AusPlots program at Adelaide University. Ben's skills include the application of spatial technologies (Remote Sensing and GIS) to environmental issues and their application in the assessment of biodiversity values, including fire mapping, resource assessments and input to development consents. He is interested in developing more objective methods for mapping arid vegetation by utilising traditional ecological clustering techniques so that the ecological information is the driving force in the image classification.

Professor **Andrew Lowe** is Chair in Plant Conservation Biology at the University of Adelaide and Head of Science within the South Australian Department of Environment and Natural Resources. He is also Director of the Australian Centre for Evolutionary Biology and Biodiversity, which integrates evolutionary biology and biodiversity science across the University, State Herbarium of South Australia and South Australian Museum. Andrew's predominant research interest is 'How do plants survive and adapt to anthropomorphised landscapes?' His group applies genomic and landscape analysis to demonstrate gene flow and selection pressure changes across a range of landscapes; contemporary, historical, fragmented and exploited.

References

Bastin G and the ACRIS Management Committee (2008). 'Rangelands 2008 – Taking the Pulse'. National Land and Water Resources Audit, Canberra, Australia.

de Preu N and Pearce D (2006). 'BounceBack Progress Report'. South Australian Department of Environment and Heritage, Adelaide, Australia.

Foulkes JN, White IA, Sparrow BS and Lowe AJ (2011). An ecological plot monitoring and stratification method for sampling Australian rangeland ecosystems; implementing AusPlots, a facility of the Terrestrial Ecosystem Research Network. Terrestrial Ecosystem Research Network, University of Adelaide, SA. <www.tern.org.au>

Graham CH and Fine PVA (2008). Phylogenetic beta diversity: linking ecological and evolutionary processes across space in time. *Ecology Letters* **11**, 1265–1277.

Godínez-Alvarez H, Herrick JE, Mattocks M, Toledo D and Van Zee J (2009). Comparison of three vegetation monitoring methods: their relative utility for ecological assessment and monitoring. *Ecological Indicators* **9**, 1001–1008.

Guerschman JP, Hill MJ, Renzullo LJ, Barrett DJ, Marks AS and Botha EJ (2009). Estimating fractional cover of photosynthetic vegetation, non-photosynthetic vegetation and bare soil in the Australian tropical savanna region upscaling the EO-1 Hyperion and MODIS sensors. *Remote Sensing of Environment* **113**, 928–945.

Johnson CN, Isaac JL and Fisher DO (2007). Rarity of a top predator triggers continent-wide collapse of mammal prey: dingoes and marsupials in Australia. *Proceedings of the Royal Society B: Biological Sciences* **274**, 341–346.

Kerle JA, Fleming MR and Foulkes JN (2007). Managing biodiversity in arid Australia: a landscape view. In *Animals of Arid Australia: Out on Their Own?* (Eds C Dickman, D Lunney and S Burgin) pp. 42–64. Royal Zoological Society of New South Wales, Mosman, Australia.

Letnic M, Tamayo B and Dickman CR (2005). The responses of mammals to La Nina (El Nino Southern Oscillation)-associated rainfall, predation and wildfire in central Australia. *Journal of Mammalogy* **86**, 689–703.

Lozupone CA and Knight R (2007). Global patterns in bacterial diversity. *Proceedings of the National Academy of Sciences of the USA* **104**, 11436–11440.

McKenzie NL, Burbidge AA, Baynes A, Brereton RN, Dickman CR, Gordon G, Gibson LA, Menkhorst PW, Robinson AC, Williams MR and Woinarski JCZ (2007). Analysis of factors implicated in the recent decline of Australia's mammal fauna. *Journal of Biogeography* **34**, 597–611.

Morton SR (1990). The impact of European settlement on the vertebrate animals of arid Australia: a conceptual model. *Proceedings of the Ecological Society of Australia* **16**, 201–213.

Pavoine S, Vela E, Gachet S, de Bélair G and Bonsall MB (2011). Linking patterns in phylogeny, traits, abiotic variables and space: a novel approach to linking environmental filtering and plant community assembly. *Journal of Ecology* **99**, 165–175.

Read J and Bowen Z (2001). Population dynamics, diet and aspects of the biology of feral cats and foxes in arid South Australia. *Wildlife Research* **28**, 195–203.

Short J and Smith A (1994). Mammal decline and recovery in Australia. *Journal of Mammalogy* **75**, 288–297.

Spaeth KE, Pierson FB, Herrick JE, Shaver PL, Pyke DA, Pellant M, Thompson D and Dayton B (2003). New proposed national resources inventory protocols on non-federal rangelands. *Journal of Soil and Water Conservation* **58**, 18A–21A.

Vittoz P and Guisan A (2007). How reliable is the monitoring of permanent vegetation plots? A test with multiple observers. *Journal of Vegetation Science* **18**, 413–422.

Watson IW and Novelly PE (2004). Making the biodiversity monitoring system sustainable: design issues for large-scale monitoring systems. *Austral Ecology* **29**, 16–30.

Watson IW, Novelly PE and Thomas PWE (2007*a*). Monitoring changes in pastoral rangelands – the Western Australian Rangeland Monitoring System (WARMS). *Rangeland Journal* **29**, 191–205.

Watson IW, Thomas PWE and Fletcher WJ (2007*b*). The first assessment, using a rangeland monitoring system, of change in shrub and tree populations across the arid shrublands of Western Australia. *Rangeland Journal* **29**, 25–37.

Whitehead P, Woinarski JCZ, Fisher A, Fensham R and Beggs K (2001). 'Developing an analytical framework for monitoring biodiversity in Australia's rangelands'. Report to the National Land and Water Resources Audit. Tropical Savannas Cooperative Research Centre, Darwin, Australia.

Wintle BA, Runge MA and Bekessy SA (2010). Allocating monitoring effort in the face of unknown unknowns. *Ecology Letters* **13**, 1325–1337.

Woinarski JCZ, Armstrong M, Brennan K, Fisher A, Griffiths AD, Hill B, Milne DJ, Palmer C, Ward S, Watson M, Winderlich S and Young S (2010). Monitoring indicates rapid and severe decline of native small mammals in Kakadu National Park, northern Australia. *Wildlife Research* **37**, 116–126.

Wynn JG and Bird MI (2008). Environmental controls on the stable carbon isotopic composition of soil organic carbon: implications for modelling the distribution of C3 and C4 plants, Australia. *Tellus B* **60**, 604–621.

DISCUSSION

22 CAN WE MAKE BIODIVERSITY MONITORING HAPPEN IN AUSTRALIA? MOVING BEYOND 'IT'S THE THOUGHT THAT COUNTS'

David Lindenmayer and Philip Gibbons

Introduction

Biodiversity monitoring has a chequered history in Australian environmental management. Almost everyone *thinks* it's a good idea, but this often does not translate to good monitoring on the ground, sound reporting of trends, improved interventions, a clear indication of the return on our investment, and ultimately better conservation outcomes. The chapter authors in this book have explored a wide range of issues associated with improving the appalling record of biodiversity monitoring in Australia. The breadth and depth of issues identified is extraordinary – which makes it extremely difficult to summarise the content of this volume. Here we focus on the major problems associated with biodiversity monitoring that were identified consistently by various chapter authors. In an attempt to ensure that this book is a constructive contribution to debates on biodiversity monitoring, we have outlined proposed solutions to the problems that were identified by chapter authors. We have enriched our treatment of problems and solutions with additional insights generated from the workshop meeting associated with this book. As indicated at the start of this book, part of the discussion of this chapter pivots around ways to help deliver on the objectives set out above and in *Australia's Biodiversity Conservation Strategy 2010–2030* (Natural Resource Management Ministerial Council 2010).

We readily acknowledge that we have not examined every one of the many important issues that were identified by the various contributors to this volume. This was unavoidable because of the limited space available in this chapter. However, it does mean that a reader can learn a lot from reading **all** of the contributions in this book – just as we have been fortunate to learn a lot from reading, editing and re-reading the chapters.

Problems and solutions

We have organised this chapter around problems and solutions collected under four key themes:

- Reasons for monitoring
- Institutionalising and funding monitoring
- Designing monitoring programs

- Technical methods for monitoring (including what to monitor and how to monitor) and using monitoring data.

Under each of these four broad themes, we discuss some of the major problems identified by the chapter authors and then counter such problems with recommended solutions (where solutions appear to exist). Notably, there are a number of cases where particular recommendations by a given chapter author contradict those made by another author. We have attempted to offer an explanation for these differences of opinion.

Reasons for monitoring

The need for better communication and the importance of monitoring

It is possible that the poor record on biodiversity monitoring in Australia (and indeed elsewhere around the world) is related, in part, to scientists and others neither advocating nor demonstrating strongly enough why it is important to do monitoring. Therefore, it seems to us that it is critical to more stridently, widely and frequently communicate the importance of monitoring (Krebs, Chapter 17) and better highlight case studies of good monitoring programs to demonstrate its value and importance. A well-designed monitoring program should deliver: (1) trends in appropriate aspects of biodiversity (e.g. populations) (Dickman and Wardle, Chapter 18); (2) an early warning of problems or threats that can be difficult, impossible and/or very expensive to reverse (Lindenmayer *et al.* 2010); (3) quantifiable evidence of conservation successes (e.g. species recovery following appropriate management intervention) (Lindenmayer, Chapter 1) as well as conservation failures (e.g. Woinarski, Chapter 2); (4) improved and more effective management interventions (Fitzsimons, Chapter 10); and (5) information on the return for our investment. One author (Gibbons, Chapter 19) noted that we have yet to quantify the cost-effectiveness of monitoring biodiversity in this country.

Communication about the importance of monitoring must come in a range of forms because traditional methods like scientific articles are only rarely read by other important partners in monitoring programs such as resource managers and policy makers (Gibbons *et al.* 2008). Several authors highlighted the importance of communicating the risks of **not** monitoring (e.g. Varcoe, Chapter 11). Finally, Fitzsimons (Chapter 10) argued there is a need to be aware that the results of monitoring programs can sometimes be misused and misquoted to support political agendas and hence scientists and other groups need to be solidly engaged in communication arising from monitoring programs.

Institutionalising and funding monitoring programs

The development of new monitoring institutions

Several chapter authors discussed, at some length, that many monitoring programs owe their success to the enthusiasm of an individual. While these champions should be recognised and supported, it was agreed that this illustrates something more sinister. A major (and perennial) problem with monitoring programs is a paucity of institutional support, a lack of long-term funding, and an absence of scientific and other expertise (Kingsford and Porter, Chapter 20; Lindenmayer, Chapter 1; Radford *et al.*, Chapter 12). These problems were recognised by almost all chapter authors and many responded with valuable suggestions to resolve them. Woinarski (Chapter 2) advocated the establishment of new institutions to coordinate monitoring, including brokering the partnerships that are fundamental to the success of monitoring programs (see below), and fostering an appropriate culture to ensure that monitoring programs are maintained in the long term. Other authors suggested that such institutions might have other roles such as storing and curating monitoring data sets or at least developing registers of data sets (Zerger and McDonald, Chapter 3), and mapping what monitoring is being done where (Garnett, Chapter 4). There is also a need for new training centres within universities to promote the development of the kinds of skills essential for good monitoring including databasing and data analysis (Garnett, Chapter 4; Freudenberger, Chapter 13; Varcoe, Chapter 11) as well as to reverse negative trends in other crucial sub-disciplines such as natural history.

Creating new institutions to support monitoring remains a true challenge in that they would need to be established in ways that do not curtail the passion of champions of monitoring programs nor limit the capacity of scientists and resource managers to be innovative, such as rapidly implementing new monitoring programs soon after major natural disturbances like bushfires (Hobbs, Chapter 5; Gibbons, Chapter 19). Similarly, new institutions designed to help maintain long-term monitoring programs are unlikely to be very effective if they are subject to the constant and high frequency restructuring that is now commonplace in natural resource management agencies Australia-wide.

Tackling problems with funding

A lack of funding is a common reason why many monitoring programs fail (Lindenmayer and Likens 2010). Moreover, there is typically a mismatch between the short time frames for funding and political cycles versus the long time frames needed for biodiversity monitoring (Hobbs, Chapter 5; Kingsford and Porter, Chapter 20). Woinarski (Chapter 2) and Lindenmayer (Chapter 1) argued the case for new kinds of funding models to overcome this hurdle such as long-term trusts and/or foundation-style support. It is also possible that more secure forms of funding for monitoring might be forthcoming if monitoring can demonstrate return on conservation investment (Gibbons, Chapter 19) and, conversely, the costs and risks of **not** doing monitoring (Lindenmayer, Chapter 1; see below).

In many respects, identifying appropriate funding models for monitoring requires research to determine what kind of approaches have been tried elsewhere in the world, including in contexts other than environmental ones (e.g. the use of proceeds from lottery revenues in the United Kingdom to support sport and the arts), and which ones are most likely to be successful in Australia (Lindenmayer, Chapter 1).

Part of the problem of ongoing funding for monitoring might be resolved if monitoring has greater political support. Many strategies are needed to generate such support. For example, the creation of a set of national environmental accounts (*sensu* Wentworth Group of Concerned Scientists 2008) would rapidly elevate the importance of biodiversity monitoring and hence support for such activities. Indeed, such accounts will only be as good (or bad) as the data that underpin them. Moreover, greater political support for biodiversity monitoring will eventuate if there is broad constituency advocating for it, including Catchment Management Authorities and other Natural Resource Management groups, the scientific community and the general public. Large-scale economic changes like those associated with the emerging 'green economy' also may provide important catalysts to generate greater political support for biodiversity monitoring. The longevity and success of monitoring programs might be increased if it is a formal or mandated requirement under international convention (Gibbons, Chapter 19; Garnett, Chapter 4), legislation (Fitzsimons, Chapter 10), or development consent (Lindenmayer, Chapter 1). Increased levels of support would aim to make biodiversity monitoring and hence biodiversity conservation a mainstream concern and not a peripheral political issue as it is currently.

Partnerships as a critical part of monitoring programs

Good monitoring programs require the integration of both good science and good management practice (Russell-Smith *et al.* 2003). Thus, while appropriate levels of funding are essential to the success of monitoring programs, partnerships among individuals with appropriate sets of skills are equally critical (Varcoe, Chapter 11, Radford *et al.*, Chapter 12). Legge and Fleming (Chapter 14) and Lindenmayer (Chapter 1) lamented the lack of engagement in monitoring by academic scientists. Such people can assist in many aspects of monitoring including experimental design, database design, data analysis, data interpretation and reporting, as well as providing guidance for citizen scientists and community groups to ensure that the data they gather are of high quality and are useful (Garnett, Chapter 4; Montambault and Groves, Chapter 15). Changes in the rewards systems for academic scientists may be needed to catalyse their involvement in monitoring programs (see Gibbons *et al.* 2008).

Natural resource managers and biodiversity managers also need to belong to the partnerships that underpin successful monitoring programs. This is to ensure that monitoring programs pass the 'test of management relevance' (*sensu* Russell-Smith *et al.* 2003) and are therefore more likely to be sustained in the long term. It is clear that successful monitoring programs will often be those that connect individuals from two or more of the following: universities, government agencies, non-government organisations and community groups (Montambault and Groves, Chapter 15).

Designing monitoring programs

Improved design

Many chapter authors raised a range of key concerns about the poor design of monitoring programs and/or have argued there are too few examples of good monitoring programs (Lindenmayer, Chapter 1), with some ecosystems and biomes in particular, being poorly monitored, such as large parts of arid and rangeland Australia (Dickman and Wardle, Chapter 18; White *et al.*, Chapter 21). Many solutions were proposed throughout this book in an effort to tackle the longstanding problems of poor monitoring design and the paucity of good monitoring programs. Possingham *et al.* (Chapter 6) eloquently outlined why it is important to recognise that not all things can be monitored and why it is critical to decide when to monitor and when not to monitor. Therefore, it is essential to ensure that monitoring programs are designed appropriately at the outset to address clearly defined problems, communicate well-articulated objectives (Robinson *et al.*, Chapter 16) and answer well-formulated questions (Lindenmayer, Chapter 1). These are well-known and fundamental components of successful monitoring that have been emphasised by many other workers (e.g. Nichols and Williams 2006; reviewed by Lindenmayer and Likens 2010). However, Gibbons (Chapter 19) noted that mandated, surveillance monitoring in which there are no *a priori* questions (see Lindenmayer and Likens 2010) can sometimes be useful in tackling environmental problems that were unforeseen at the commencement of a project. The impact of pesticides on eggshell thinning in birds in North America is a classic example – a finding that corresponds with the insights of Hobbs (Chapter 5) who noted that monitoring data are often useful in a range of different contexts beyond a monitoring one. This highlights the challenge articulated by Lindenmayer (Chapter 1) and Zerger and McDonald (Chapter 3) about the need to find new ways to better link data and insights from question-driven and mandated monitoring programs.

A significant problem in the design and implementation of monitoring programs is that a prolonged period may elapse between a given management intervention and the response of some elements of biodiversity. A number of strategies can help deal with this 'slow feedback' loop including using cross-sectional approaches in monitoring design to generate rapid initial insights (Lindenmayer and Likens 2010), identifying indicators that respond rapidly to management, and employing predictive modelling to provide an early indication of likely outcomes and formulate additional hypotheses for testing (Jones *et al.* 2011; Ferrier, Chapter 7).

Although there are many problems associated with the design of biodiversity monitoring programs, there are nevertheless examples of good ones and it is critical to build on them (Krebs, Chapter 17; Kingsford and Porter, Chapter 20). The value of these existing programs often increases over time and hence there are issues with how to maintain them, as well as ensure succession planning when the champions for such programs retire or die (see below).

Finally, there may be value in completing meta-analyses and systematic reviews to determine what needs to be monitored and what kinds of monitoring programs have worked best to avoid repeating past mistakes in establishing new monitoring programs (Montambault and Groves, Chapter 15). There also may be value in establishing pilot programs to determine the most appropriate designs and likely effectiveness of new monitoring programs (Varcoe, Chapter 11). Similarly, the design of monitoring programs might be improved if they embrace some business approaches such as auditing expenditure and developing metrics that highlight return on investment (including determining when monitoring programs cease to be effective).

Resolving arguments about what to monitor

Another longstanding problem with biodiversity monitoring programs has been the protracted (and often bitter and unresolved) debates about what to monitor (Gardner 2010). Several chapter authors in the book offered important 'antidotes' to this chronic 'disease' plaguing monitoring programs. For example, Krebs (Chapter 17) highlighted the importance of prioritising the entities to be targeted for monitoring as it is 'impossible to monitor everything'. It was further suggested that such entities should be those which tell us about the condition, function and integrity of a given system (Hobbs, Chapter 5) and which help answer key questions of management relevance (Montambault and Groves, Chapter 15), thereby ensuring there is a social mandate for what is being monitored and how monitoring is designed (Burgman *et al.*, Chapter 8). Notably, Legge and Fleming (Chapter 14) emphasised how the development of appropriate questions of management relevance and the selection of appropriate entities to be measured in monitoring programs need to be informed by a detailed understanding of the ecosystem of interest.

Some chapter authors stressed the critical importance of monitoring management inputs as well as biodiversity outcomes. Gibbons (Chapter 19) noted how rarely monitoring programs gathered data on the management practices undertaken at monitoring sites (e.g. the grazing regime, burning regime, or level of pest control) or reported how much financial investment there had been in management interventions at such sites. Yet, such data are critical for linking the response of biodiversity to management practices and the cost-effectiveness of management interventions. Conversely, Hobbs (Chapter 5) highlighted the importance of gathering data on biodiversity responses and not solely crude estimates management inputs (e.g. kilometres of fencing) as occurred in some infamous cases like the Natural Heritage Trust.

Improving monitoring through reviews and embracing Adaptive Monitoring

Even when robust monitoring programs are based on good questions of management relevance and target an appropriate subset of entities for measurement, there is still a need to subject them to regular review (White *et al.*, Chapter 21; Possingham *et al.*, Chapter 6). This is essential to ensure they can be improved, for example, by adding new entities to be measured as a result of new insights that have been gained (Krebs, Chapter 17) or when new questions of management relevance become important. Such a cycle of continuous improvement in monitoring programs has been termed 'Adaptive Monitoring' (Lindenmayer and Likens 2009) in which a monitoring program can evolve and develop in response to new information, new questions or to improve obsolete approaches and monitoring protocols. A critical aspect of Adaptive Monitoring is the need to ensure that a change of monitoring protocols does not breach the integrity of the long-term nature of the data record. Calibration of different field methods can help solve this potential problem (Lindenmayer and Likens 2010).

Improved data standards

Several chapter authors noted there are many disparate kinds of biodiversity monitoring programs around Australia that are characterised by marked differences in experimental or survey design, field protocols, entities targeted for measurement, and spatial and temporal scale of implementation. Such differences have led to a lack of national data sets to quantify national trends, although the Atlas of Australian Birds is one of a handful of notable exceptions (Garnett, Chapter 4). Given this, some chapter authors called for national standards for monitoring programs and data gathering (Gibbons, Chapter 19). There are some useful guides for national biodiversity monitoring data standards that can be drawn from other areas of environmental monitoring, such as the water standard that has been developed by the Bureau of Meteorology (Vardon, Chapter 9). Notably, any standards that are developed would not remain static but evolve over time with new technology.

While data standards are undoubtedly important, especially for the validity of subsequent analyses and interpretation of monitoring data, it is inappropriate to demand that precisely the same full suite of entities should be gathered in all monitoring programs across Australia. There are many obvious reasons for this. The principal one is that entities (such as particular groups of biota) that are sensible for measurement in, for example, the wet forests of south-western Tasmania will be markedly different than those entities appropriate in the tropical savannas of northern Australia. Counts of butterflies could be valuable for answering particular questions in tropical rainforests but of limited use in desert environments. Moreover, different problems, different questions of management relevance and different biota in different ecosystems mean that a diversity of kinds of monitoring programs will be needed to match monitoring design with monitoring objectives (Freudenberger, Chapter 13). Standards should therefore generally focus on higher-order issues such as statistical design, metadata and data access, rather than finer levels of detail (e.g. indicators that must be measured) (Gibbons, Chapter 19).

Technical methods for monitoring (including what to monitor and how to monitor) and using monitoring data

It is not possible to discuss monitoring programs, nor write about them, without at least some technical dialogue about what to monitor, how to monitor and how to use monitoring data. Almost all chapter authors touched on these issues and a range of interesting recommendations were made. Hobbs (Chapter 5) argued that monitoring data should be gathered in ways that facilitate later use by others. This resonates with the suggestions of Krebs (Chapter 17) that the methods used in field data collection should be carefully documented and published so that others can repeat data collection protocols at a later date if needed. Moreover, there is a need for high quality data management to facilitate effective and timely use of data for informed decision-making (Garnett, Chapter 4), such as soon after major natural disturbance events like those associated with the 2009 bushfires in Victoria (Gibbons, Chapter 19). This is particularly true for large monitoring data sets that have been gathered over a prolonged period and where the champion for that program aims to retire but has a desire for the program to continue and/ or wants information to be used by others in the future (Kingsford and Porter, Chapter 20). This highlights the potential coordinating and succession planning roles of new monitoring institutions, as outlined above (Zerger and McDonald, Chapter 3).

The importance of new technology

Several chapter authors discussed the importance of new technologies for promoting the rapid collection of large amounts of field data from monitoring programs. For example, Dickman and Wardle (Chapter 18) emphasised the value of remotely sensed data sets as an adjunct to other major data sets and, in turn, their value for explaining long-term patterns in animal abundance in desert landscapes. Similarly, Zerger and McDonald (Chapter 3), Dickman and Wardle (Chapter 18), White *et al.* (Chapter 21) and Fitzsimons (Chapter 10) highlighted the potential of a wide range of new technologies that can make data collection more cost-effective or facilitate greater public involvement in monitoring.

There is also a need to connect on-ground monitoring methods and data collection with other large-scaled approaches like remote sensing (Ferrier, Chapter 7) and there are now increasing numbers of examples of this in an Australian context (e.g. Youngentob *et al.* 2011). However, it is crucial to link these approaches through well-articulated objectives and posing good scientific questions, otherwise there is a risk of being overwhelmed in a 'sea of data' and failing to make critical discoveries. Shanklin (2010) provides an example of this problem in studies of the ozone layer.

Improving the availability of monitoring data

Several chapter authors made calls to make all monitoring data publicly available (e.g. Fitzsimons, Chapter 10). This is an important point as monitoring data sets have often been gathered through public funding. However, an important issue is the maintenance of intellectual property and the development of new ways to allow scientists and others who have gathered data to be able to publish them in a timely way before they are made available publicly. Resolution of intellectual property and data licensing issues should help reduce instances of misuse of data. There also may be value in examining new kinds of reward systems that acknowledge the contributions of scientists and others who have collected data sets that are used by others. For example, developing new kinds of metrics which reflect the amount of use of a given data set by others and which underpin rewards for levels of data usage. Similarly, there is a need for the editors of journals to demand that the authors of articles acknowledge the sources of data, perhaps through an explicit statement of origin of data sets using approaches akin to ethics approval statements that are now common in many publications.

General recommendations

Given the extent of thought that underpins the chapter contributions in this book, coupled with the extensive discussions in the workshop that preceded the publication of this book, we believe it would be remiss not to provide three broad recommendations.

1. Biodiversity monitoring is chronically under-funded and this needs to be urgently rectified. The expenditure on environmental management in Australia exceeds $12 billion

annually (Beeton *et al.* 2006). We believe that at least 10 per cent of this should be dedicated to biodiversity monitoring, and in turn, guiding better environmental management. Further funding for biodiversity monitoring programs could be derived from monies quarantined from development projects. There is also a need for dedicated research explicitly aimed at identifying the most appropriate ways to sustain long-term funding for biodiversity monitoring programs.

2. There is a critical need for new institutions to support biodiversity monitoring in Australia.

   a) First, there is an urgent need to create a coordinating institution with many roles such as: providing funding assistance to instigate and maintain long-term monitoring programs, curating long-term data sets, developing national standards for monitoring data and monitoring programs, brokering partnerships between individuals and organisations, and fostering succession planning in existing long-term programs.

   b) Second, there is a need for a new research and training institution to promote the development of new generations of people with the kinds of skills essential to establish and maintain long-term biodiversity monitoring programs, to communicate the need and value of biodiversity monitoring programs, and to foster innovation and continuous improvement in monitoring programs.

   The establishment and maintenance costs of these institutions could be readily supported through the 10 per cent of ~$12+ billion spent annually on environmental management and re-directed to biodiversity monitoring.

3. There is a need to identify new ways to better communicate the importance of biodiversity conservation to politicians and society in general. One approach is to better connect the need for biodiversity conservation with improved human population health (Chivian and Bernstein 2008). Borrowing from a recent editorial in *Science* on science-based health care, 'a traditional misconception is that spending on health is a social burden, instead of being a strategic investment essential for each nation's socioeconomic development' (Zhu 2010). This argument is equally valid for science-based environmental monitoring management, conservation and the protection of biodiversity. Indeed, recent work (e.g. European Communities 2008) shows the economic and employment benefits of appropriate environmental investment to be substantial while the costs of inaction are significant.

Concluding comments – moving beyond 'It's the thought that counts'

This book contains many suggestions for moving biodiversity monitoring beyond 'It's the thought that counts'. It is clear that we need more monitoring, for additional monitoring to be far better than it generally has been to date, better training of those involved in monitoring programs, and new and different kinds of institutions to coordinate, implement and maintain monitoring programs. Our sincere hope is that many of the insightful suggestions made in this book might be adopted by policy makers, resource managers and scientists, and that the importance of doing good monitoring and then reporting and acting on monitoring outcomes is recognised. We believe this will ultimately lead to better outcomes for biodiversity in this country.

References

Beeton RJS, Buckley KI, Jones GJ, Morgan D, Reichelt RE and Trewin D (2006). 'Australia state of the environment'. Independent report to the Australian Government Minister for the Environment and Heritage. Department of the Environment and Heritage, Canberra.

<http://www.environment.gov.au/soe/2006/publications/report/pubs/soe-2006-report. pdf>

Chivian E and Bernstein A (2008). *Sustaining Life. How Human Health Depends on Biodiversity*. Oxford University Press, Oxford, England.

European Communities (2008). 'The economics of ecosystems and biodiversity'. European Communities, Wesseling, Germany.

Gardner T (2010). *Monitoring Forest Biodiversity. Improving Conservation Through Ecologically Responsible Management*. Earthscan, London.

Gibbons P, Zammit C, Youngentob K, Possingham HP, Lindenmayer DB, Bekessy S, Burgman M, Colyvan M, Considine M, Felton A, Hobbs RJ, Hurley K, McAlpine C, McCarthy MA, Moore J, Robinson D, Salt D and Wintle B (2008). Some practical suggestions for improving engagement between researchers and policy-makers in natural resource management. *Ecological Management & Restoration* **9**, 182–186.

Jones JPG, Collen B, Atkinson G, Baxter PW, Bubb P, Illian JB, Katzner TE, Keane A, Loh J, McDonald-Madden E, Nicholson E, Pereira HM, Possingham HP, Pullin AS, Rodrigues AS, Ruiz-Gutierrez V, Sommerville M and Milner-Gulland EJ (2011). The why, what, and how of global biodiversity indicators beyond the 2010 target. *Conservation Biology* **25**, 450–457.

Lindenmayer DB and Likens GE (2009). Adaptive monitoring: a new paradigm for long-term research and monitoring. *Trends in Ecology and Evolution* **24**, 482–486.

Lindenmayer DB and Likens GE (2010). *Effective Ecological Monitoring*. CSIRO Publishing, Melbourne.

Lindenmayer DB, Likens GE, Krebs CJ and Hobbs RJ (2010). Improved probability of detection of ecological 'surprises'. *Proceedings of the National Academy of Sciences of the USA* **107**, 21957–21962.

Natural Resource Management Ministerial Council (2010). 'Australia's Biodiversity Conservation Strategy 2010–2030'. Australian Government Department of Sustainability, Environment, Water, Population and Communities, Canberra. <http://www.environment. gov.au/biodiversity/strategy/index.html>

Nichols JD and Williams BK (2006). Monitoring for conservation. *Trends in Ecology and Evolution* **21**, 668–673.

Russell-Smith J, Whitehead PJ, Cook GD and Hoare JL (2003). Response of *Eucalyptus*-dominated savanna to frequent fires: lessons from Munmarlary 1973–1996. *Ecological Monographs* **73**, 349–375.

Shanklin J (2010). Reflections on the ozone hole. *Nature* **465**, 34–35.

Wentworth Group of Concerned Scientists (2008). 'Accounting for nature. A model for building the National Environmental Accounts of Australia'. Wentworth Group of Concerned Scientists, Sydney.

Youngentob KN, Roberts DA, Held AH, Dennison PE, Jia X and Lindenmayer DB (2011). Mapping two Eucalyptus subgenera using multiple endmember spectral mixture analysis and continuum-removed imaging spectrometry data. *Remote Sensing of Environment* **115**, 1115–1128.

Zhu C (2010). Science-based health care. *Science* **327**, 1429.

Appendix 1: the workshop design, data capture and recording process

One of the main challenges confronting a book with multiple contributors is how to capture everyone's best thinking and ideas, and then to have a process in which everyone can agree on the key ideas. This was one of the most important design challenges in bringing all contributors together for a two-day workshop at ANU – how to effectively collaborate and do real work together?

The traditional workshop process typically involves flip charts and whiteboards as the primary capture tools. Sometimes scribes are employed to try to capture in real time the key points emerging from the conversations.

Rather than take the traditional workshop approach, the lead authors decided to use a collaborative web-based decision-making tool called iMEET!, developed by Global Learning, a Canberra-based consultancy group. iMEET! is purpose-built to support the rapid capture of individuals' ideas or comments. Once in the iMEET! system, ideas are able to be structured by category, tagged by key word, ranked and voted on in a process that supports a group's capacity to quickly agree on key areas for action.

Before the workshop, all delegates were asked to log in to the purpose-built iMEET! workshop website and upload their draft chapters and workshop presentations. Eighteen chapters and 12 presentations were uploaded before the workshop.

Based on the draft chapters, a shortlist of key lessons and categories was developed by the workshop facilitators. These data provided valuable input into both the workshop design and the development of six broad categories to help collate the data collected during the workshop.

The iMEET! technology utilised 12 iMAC computers, all wirelessly connected to an iMEET! web server, thereby enabling small groups of 2–3 delegates to contribute their comments and ideas via keyboard into a shared topic database. As ideas were contributed, they were assigned to one of six categories as well as being tagged as an observation, solution or problem. All contributions were projected in real time via data projector to a large screen.

With all ideas visible, facilitated large group discussion identified patterns in the data, questions arising, or anything missing. New entries or editing of existing entries were keyed in and instantly visible to all, as needed.

The workshop agenda was specifically designed to provide maximum time for delegates to both talk and contribute their best thinking to specific topic areas as they were raised. The key idea was to not have the technology dominate, but serve as a productive tool to capture the best of the large group and small group conversations. The opportunity cost of getting the group together meant that the collective and individual wisdom needed to be thoughtfully 'harvested'.

Over the duration of the two-day workshop there were four 'speedtalk' sessions. These were short 5-minute summary presentations by authors on the key points of their chapter and areas of expertise. At the end of each 'speedtalk', all delegates contributed their comments into iMEET! These ideas were categorised and word-tagged on entry, enabling easy sorting and display of the data. Over the two days, a total of 406 individual data entries were captured in iMEET!, totalling more than 8000 words.

At the end of day one, the workshop facilitators undertook a review of the contributions. Day two commenced with the presentation of a summary to the delegates of approximately 40 key problems and solutions that had emerged. After group discussion and refinement of the list, each delegate was asked to nominate (via an iMEET! voting process) their top 10 items from the list. The first round of voting helped eliminate nearly 50 per cent of the items, either through combining with other items or not being deemed of high priority. A second vote settled on 15 agreed key problem and solution areas. A facilitated large group session then discussed and identified solutions to most of these key areas. As solutions were agreed these were keyed directly into the relevant iMEET! section, at the same time being projected on the large screen.

To complete the workshop, all the data were published online with access for delegates to review. All data from the workshop were also exported from the website as an Excel spreadsheet to enable further analysis and documentation, as needed.

In summary, the key iMEET! processes included:

Pre workshop
- The capacity to have a dedicated workshop website that enabled authors to easily upload and share their chapters and presentations with all delegates.
- The development of an agenda linked to interactive sessions.

During the workshop
- All delegates contributing, simultaneously, comments on all workshop presentations. These comments were sorted by contributors across six category areas as well as being tagged as a problem, solution or observation.
- From a refined list of problems and solutions, a voting process that led to an agreed shortlist of key problems and solutions.

Post workshop
- Publishing of the data online.
- Export of all data as an Excel spreadsheet for processing.

Reference
http://imeet.com.au

Index

Aboriginal people, monitoring and   161
    see also Indigenous Protected Areas;
        Indigenous ranger programs
academic journal articles   38–9, 96, 154, 194
academics
    conservation and   117, 135
    failure of   129–30, 131
Access database   153
accountability   15–22, 25, 145
acid rain phenomenon   53
action fund   136
Actions for Biodiversity Conservation
    program   107
activity monitoring   50, 51, 67, 106
adaptive management   19, 103, 105, 106, 122,
    124, 134, 136, 137, 165–70, 183, 198
Advanced Very High Resolution Radiometer
    (AVHRR)   24
aerial surveys   50, 172
Agreement on the Conservation of Albatrosses
    and Petrels   34, 37
agri-environment program (England)   143
aims, ecological monitoring   50–3
Alaska Highway   152
albatrosses   37
Alberta Biodiversity Monitoring program   155
alpine ecosystems   102
Alpine National Park (Victoria)   92, 93, 97
Antarctic Division   37
arid Australia   157–64, 179, 196, 199
Arid Recovery project   160, 161, 187
assessment endpoints   63–70, 72
Atlas of Australian Birds   24, 26, 29, 38, 39, 82,
    93, 94, 95, 96, 130, 142, 144, 198
Atlas of Living Australia (ALA)   23, 24, 25, 28,
    96, 186
auditing, need for   124
AuScope   27
AusCover products   182, 184
AusPlots Rangelands surveys   162, 180–2,
    183–5, 186–7
Australia
    conservation strategies   122
    ecological monitoring in   43–8, 171–7
    mega-diversity and   1
Australian Alps   92

*Australian Biodiversity Conservation Strategy
    2010–2030*   1–4, 20, 80, 122, 193
Australian birds   33–41
Australian Bureau of Statistics (ABS) 17, 24, 26,
    28, 38, 79, 81, 82–4, 85, 129
Australian Collaborative Rangelands
    Information System (ACRIS)   181, 187
Australian Desert Expeditions   162
Australian Geographic   162
Australian Government   34, 143, 176, 181
Australian National Audit Office (ANAO)   1, 80
Australian National University   1
Australian Ocean Data Network   27
Australian Research Council   28, 40, 58, 117,
    163
Australian Spatial Data Directory (ASDD)   27,
    168
Australian Wader Studies Group (AWSG)   34,
    36, 38
Australian Water Account   81
Australian Wildlife Conservancy (AWC)   9, 111,
    127–32, 160, 161

back loop   165–70
Barrow Island   20
baseline surveillance monitoring program   180,
    183
Bennett, Andrew   113
'big-picture' change   65
bilby (*Macrotis lagotis*)   160
biodiversity decline   129
Biodiversity Prediction using Ecological
    Processes (BioPrEP)   114
Biological Surveys Committee (BSC)   161
bioregions   181
biotic data   159, 171–6
Birdata   94
BirdLife International   35, 36
Birds Australia   34, 35, 38, 82, 117, 122, 130, 141,
    142, 144, 162
bird surveys   33–41, 44, 112, 113, 162
Bismarck Sea communities   134
Bogong High Plains   92
Bonn Convention on Migratory Species   34
Booderee National Park   8
boom and bust cycles   113, 158, 160, 173

bottom-up (field-based) model   65–6, 67, 124, 128, 138
box-gum grassy woodlands   9
box-ironbark forests   113
British Trust for Ornithology   38
brown stringybark (*Eucalyptus baxteri*)   92
budget monitoring   45–6, 174, 175–6
buffel grass (*Cenchrus ciliaris*)   159
buloke tree (*Allocasuarina Luehmannii*)   92, 93
Bureau of Meterology (BoM)   81, 23, 27, 28, 144, 198
Burned Area Emergency Response program   169
Bush Blitz   187
bushfire management   16–17, 19, 93, 103, 112, 116, 130, 131, 159, 165–70
Bush Heritage Australia   9, 111–19, 158, 160, 161, 163
Bushtender   85
business model   19, 145
butterfly counts   198

calibration   73
camel treks   162
camera traps   161
Canada   151–5
cane toad (*Bufo marinus*)   159
Capital Region database (Greening Australia)   124
captive breeding   50
Caring for our Country (CfoC)   28, 141, 143
Catchment Management Authorities   9, 195
cattle grazing (Alpine National Park)   92, 93, 97
Census and Statistics Act 1905   81
central data repository   168
Central Highlands (Victoria)   85, 102
Centre for Evidence-based Conservation   136
centre pivot irrigation   92
change tracking (biota)   159
Channel Country   160
charitable-status organisations   121
China–Australia Migratory Bird Agreement (CAMBA)   34
citizen science   26, 29, 94, 96, 134, 141, 144
climate change
    Australian birds and   142
    Canada   153
climatic data   166, 182, 186
coarse-level tracking   158–9
Collaboration for Environmental Evidence   25, 136
collaborative monitoring partnerships   108

Collaborative Rangelands System   27
Commonwealth Environmental Research Facility   58
Commonwealth Scientific and Industrial Research Organisation (CSIRO)   36, 82, 124, 174
Community Access to Spatial Information (CANRI)   168
community dynamics studies   151
community involvement   52–3, 134, 141–7, 194, 196
conservation management   116–17, 135, 136, 158
Conservation Measures Partnership (CMP)   133, 134
control charts   74
Convention on Biological Diversity (CBD)   33, 34, 65
Convention on the Conservation of Migratory Species of Wild Animals (CMS/Bonn Convention)   34
Convention on Wetlands of International Importance (Ramsar Convention)   34
coordination, need for   28
core data   169
cost, aerial surveying and   172
cost analysis, failure of   103–4
cost–benefit data   50, 54, 55–6, 169
cost-effectiveness   157, 161, 167, 173
    see also budget monitoring
covenants   94
Cravens Peak Station   114
cross-cutting analysis   27

dam flows   134
data accessibility   46, 83–4, 96, 175
data collection   52, 175, 198–9
    inadequacy of   27–8, 94, 95, 96
data integration   28–9, 84–5
data management   36, 85, 169, 175, 198–9
DataONE   24
data sets   144, 166–7, 168, 173, 195
Deakin University   113
decision-making   65, 73, 116
Department of Environment and Conservation   45
Department of Environment, Climate Change and Water   124
Department of the Environment, Water, Heritage and the Arts   58
Department of Sustainability and Environment (DSE)   103, 107

Department of Sustainability, Environment,
    Water, Population and Communities   82
Department of the Environment and Water
    Resources   97
Desert Ecology Research Group   112, 117, 157,
    161, 162
desert stringybark (*Eucalyptus arenacea*)   92
dessication resistance   158
Dickman, Chris   112
diversity   64–5
DNA barcoding   184, 185, 186, 187
dormancy   158
duck hunting   171, 172, 173

early warning systems (ecological)   194
Earthwatch   117, 162
East Asian Australian Flyway   36
East Australian Waterbird Survey
    (EAWS)   172–7
eastern bristlebird   8
Eco-informatics   23–31, 182, 186
Ecological Integrity (EI) monitoring   102, 104
ecological monitoring   43–8, 49–61, 143–5
ecological variation   112–13, 115
e-commerce   29
economic motivations   162
Education Infrastructure Fund Super Science
    Initiative   181
education programs   52
effectiveness monitoring   8, 17–18, 106–7
eggshell thinning   196
electronic data acquisition   29, 153
endpoint hierarchies   71–8
environmental degradation   85–6, 171
environmental management   15–22, 25, 71–8,
    79–88, 91–9, 107, 161
error checks   153
Ethabuka Station   114
evaluation culture   105–6
event censuses   163
Excel spreadsheet   203
extinction probability   54

farmland bird species survey (England)   144
fauna abundance   28
Fawcett, Maisie   92
feedback   73, 197
feral camel (*Camelus dromedarius*)   160
feral cat (*Felis catus*)   159
feral predator control   116
field data   28, 124, 130, 153, 154
First Nations people   154

fishing regulations   19
fish populations   50, 134, 173
flood peaks   83
flying surveys   174
flyways   36
food webs   151–2
fore loop (adaptive cycle)   166
forest management data   167
Framework Convention on Climate Change
    National Greenhouse Gas Inventories
    (UN)   24
freshwater ecosystems   171, 174
funding   10–11, 136, 162, 197
    Birds Australia   144
    community organisations   143, 145
    shortfalls of   195–6, 199–200
funding cycles   130, 186

genetic assessment   185
Geoscience Australia   27
Global Index of Vegetation Plot Databases   24
Global Learning   202
GloVis service   26
Gondwana Link   123, 135
Google   29
Gorgon development   20
Gov 2.0   27
government spending   80
government–NGO partnerships   38, 39
GPS   187
grassland systems   44
green economy   195
Green River (Kentucky)   134
Greening Australia   121–6
grey-crowned babbler (*Pomatostomus
    temporalis*)   8, 142
ground berry production   152
ground-level monitoring   158, 161–2
Group on Earth Observations Biodiversity
    Observation Network (GEO BON)   24, 25
guilds   113, 173

habitat monitoring   142, 143, 157, 160
hairy-footed dunnart (*Sminthopsis hirtipes*)   158
handbook production   153
harvestable animals   50
Harvey, Judith   123
help-desk   137
hierarchical spatial structuring   106–7, 116–17
high conservation value land areas   113
historic data   44, 86
Hobart   37

human population health   200
hydrology   81, 173

iMEET!   3, 202–3
impact monitoring   123
Important Bird Areas (IBAs)   34, 35–6, 38
incidental observation recording   106
Indigenous Protected Areas   36
Indigenous ranger programs   187
individual inputs   122–3, 128, 141
information technology   94, 96
innovation   195
insectivores   113
INSPIRE Directive   28
institutional investment   124
institutional leadership   27
institutionalised support   11, 115, 200
integrated framework   67–8
Integrated Marine Observing System
    (IMOS)   23
Intensive Landuse Zones   123
Interim Biogeographic Regionalisation of
    Australia (IBRA)   123
international agreements   33
International Paper project   137
International Plan of Action for Reducing
    Incidental Catch of Seabirds in Longline
    Fisheries   34
International Recommendations for Water
    Statistics   81
International Union for the Conservation of
    Nature (IUCN)   33–4
intervention policies   122
invertebrates   158
investment   168, 175, 197
irrigation areas   83
issue awareness raising   51–2

Japan–Australia Migratory Bird Agreement
    (JAMBA)   34

Kakadu National Park   16–17, 19
kangaroo harvesting   50, 162
Kenya   35
Kluane Lake ecosystem (Yukon)   151–5
Knowledge Network for Biocomplexity   24
Kruger National Park   131
Kyoto Protocol   24

land clearance   129
land management data   162, 168
land purchases   111

Landsat satellites   24, 26, 158–9
landscape disturbance   74
landscape health   115
large areas monitoring   64
large-scale surveys   173
Leaf Area Index   184
leaf samples   186
legislation, data collection and   96
Lindenmayer, David   85
livestock removal   160
Living Murray program   102, 143
Living Planet Index   20
local communities, empowerment of   35–6
Long Term Ecological Research (LTER)   183
long-term data sets   44, 173, 174, 175, 199, 200
long-term funding   20, 194
long-term monitoring   51–2, 92, 113, 130, 131,
    158, 163, 166, 195

mallee   102, 112
Management Effectiveness Framework
    (WCPA)   107
management inputs   167
management outcomes   127–32
management systems   45, 53, 55, 65, 72, 102–3,
    194, 196, 197–8
mandated surveillance monitoring   11–12, 196
marine protected areas (MPAs)   134, 136
marine turtles   20
Maron, Martine   92–3
master business plan   133
Max Planck Society   128
measurement standardisation   72–3, 153, 169
Measures Summits   133
meta-analysis   136, 169, 197
migratory birds   35, 37, 38
Millennium Developmental Goals   33
mobile desert-dwellers   158
Moderate Resolution Imaging Spectroradiometer
    (MODIS)   24, 186
monitoring advocacy   37
monitoring costs   55
monitoring failures   64
Monitoring for Improved Biodiversity Outcomes
    conference   118
monitoring program design   1–4, 7–13, 18–19,
    115–16, 107, 154, 168
monitoring skills   195
monitoring staff, need for   36–7
monitoring studies, absence of   9–10
monitoring, evaluation, reporting and
    improvement (MERI) plan   143, 145

Montana Legacy Project   137
montane forest   112
multi-criteria monitoring, primary-objective
   monitoring versus   56–7
Murray–Darling Basin   174
mushroom production   152, 154

National Biodiversity Strategy   65
National Carbon Accounting System
   (NCAS)   23, 24, 26
National Collaborative Research Infrastructure
   Strategy (NCRIS)   23, 24, 181
national coordinating facility, need for   19–20
national data collection standards   168–9, 195
National Heritage Management System   102
National Land and Water Resources Audit
   (NLWRA)   26–7, 82, 129
National Parks (Canada)   153
National Parks Service (NPS) (US)   105, 107
National Plan for Environmental Information
   (NPEI)   23, 82
National Science Foundation (US)   28
National Vegetation Information System
   (NVIS)   27, 187
native rodents   159
native species, threats to   10, 16–17, 159
Native Vegetation Inventory Assessment   129
Natural Heritage Trust (NHT)   10, 26, 45, 80,
   123, 197
natural resource management (NRM)   67, 122,
   141–7, 195
Nature Conservancy   133–40
   Australia Program   135
Nature Foundation South Australia   111
New Ireland   134
New South Wales   9, 173, 174
New Zealand   102
newspaper publicity   153
Nobel Prize for Economic Sciences   80
non-government organisations   38, 39, 127
Northern Territory   157
northern Victoria   142
not-for-profit organisations   111–19
null hypothesis tests   73–4

objectives, definition of   142–3, 173–4
observation technologies   68
Office of Environment and Heritage
   (Victoria)   102
omnibus surveillance monitoring   166
on-ground measurement   85

Open Standards for the Practice of
   Conservation   133
Operation BounceBack   187
optimal harvesting   54–5
outcomes monitoring, activities monitoring
   versus   67, 104–5
output measures   129
overgrazing   162, 182, 183

Park Grass Experiment   44
parks management   101–10, 154
Partnership for the Conservation of Migratory
   Waterbirds (Flyway Partnership)   34
partnerships   135, 183, 196
peer-reviewed articles   135
performance measurement   37–8
personal motivation   162
personnel training   175
pest species   159
pesticides   53
petrels   37
pilot projects   81, 134–5
plant collections   162
plot assessment   180, 181, 183, 187
point intercept method   187
poison baiting   160
policy actualisation   193–201
political aims   93
politics, funding and   195
prediction testing   57–8
pre-planned crisis responses, need for   169
primary data acquisition, limitations of   28
primary-objective monitoring   56–7
private conservation estate   94, 111–19
program evaluation   24, 133–40, 196, 197
Project Review Committee (Nature
   Conservancy)   137
project teams   135, 136
Protected Area agencies (Canada)   104
Protected Areas (Victoria)   101, 102, 108, 174
protocol-based monitoring   106
public availability of data   199
public awareness raising   52–3, 85–6

quality assurance process   73
quantitative data   54, 186, 194
question definition   116

rainfall patterns (arid areas)   159
rainforest   112
Ramsar sites   35, 176

rangelands Australia   179–90, 196
rapid large-scale surveys   173
red fox (*Vulpes vulpes*)   8, 159
Red List Index   34, 38–9
Reef Watch   144
refuge habitat   114
regional report cards   74–5
Remote Economic Participation (Desert Knowledge) Cooperative Research Centre   187
remote sensing   85, 158–9, 162, 186, 199
reptile records   162
Republic of Korea–Australia Migratory Bird Agreement (ROKAMBA)   34
research and training institution   200
research monitoring   24, 143, 145
Research Partners Program (Victoria)   104, 108
research (arid desert)   158
researchers   39, 160–1, 162
restoration strategies   121–6
results monitoring   123
return on investment   49–61, 194
river flows   173, 174
river red gum (*Eucalyptus camaldulensis*)   92
Rothamsted (England)   44
Royal Commission on Black Saturday Fires 2009   165–70, 199
    see also bushfire management
Royal Geographical Society of Queensland   162
rufous scrub-bird (*Atichornis rufescens*)   63

sanctuary management   127–9
sand dune habitat   157, 159
satellite imagery   24, 26, 28, 68, 166, 173, 182
school groups   153
Science   200
science, data collecting and   129, 130
seasonal variability   183
self-paced online training   136
senior management   135, 137
senior researchers, need for   153
serendipitous discoveries   53, 56, 57
shifting baselines   51
Shorebirds 2020 project   142
short-term cyclic funding   18
Signs of Healthy Parks program   102–3, 105, 106–7
Simpson Desert   157, 158, 159, 160, 161, 162
single species monitoring   115
site scarcity   160, 161, 162
site selection   47, 115, 142

site-based field data   94, 130, 162
Skagit River Basin project   136
skill maximisation   196
slow feedback loop   197
small mammal populations   151–3
snowshoe hare (*Lepus americanus*)   152
social involvement   72, 96–7
software   133
soil seed bank   158
soils   27, 185
South Africa   131
South Australia   160, 173
south-eastern red-tailed black cockatoo (*Calyptorhynchus banksii graptogyne*)   92, 93
southern cassowary (*Casuarius casuarius johnsonii*)   37
south-west Tasmania   198
spatial variability   182
species extinction   157
species monitoring   158, 160, 187
speckled warbler (*Pyrrholaemus sagittatus*)   113
speedtalk sessions   202
spinifex (*Triodia basedowii*)   159, 160
spruce bark beetle   153
standard lexicon   124
State of the Environment report (Victoria)   93
State of the Parks program   102
Statewide Land Cover and Trees Study (SLATS)   24, 26
Statistics Canada   79
status monitoring   122
Stilt   36
strategy development   37–8
strategy effectiveness monitoring   122, 123
stratification   116–17, 182
stream gauges   28
succession planning   175
superb fairy-wren (*Malurus cyaneus*)   113–14
supplementary monitoring fund   136
'surprise' data findings   44
surveillance monitoring   49, 93–4, 95, 106, 166, 167, 197
survey bands   172
Sustainable Rivers Project   134, 136
System of Environmental-Economic Accounting (SEEA)   79, 81
System of National Accounts   80, 82
system-driving events   47

tallgrass prairie restoration   135
targeted monitoring   159, 183, 197

Tasmanian devil (*Sarcophilus harrisii*)   56
Tasmanian Government   34
Tasmanian Land Conservancy   111
Tattler, The   36
taxonomic identification   185
technological developments (monitoring)   154, 186
TERN AusPlots   180, 181–2, 186
    see also AusPlots
terrestrial animal dynamics   152–3
Terrestrial Ecosystem Research Network (TERN)   9, 10, 23, 24, 28, 44, 162
Thomas Foundation   12
threatened species   17, 54, 129, 143, 145
three-dimensional structural information   186
timber land purchase   137
time frames   154, 195
top-down management model   65–6, 67, 138
tropical northern regions   179
tropical rainforests   112, 198
Trust for Nature   141, 145
tussock grasslands (New Zealand)   7

Uluru National Park   161
United Nations   24
United States of America (USA)   134, 169
University of New South Wales   174
University of Sydney   112, 114, 157
untargeted monitoring   53

validation   73
vegetation cover   28, 74, 75, 158, 159, 184, 185, 186, 187

vertebrate species   158, 159
vested interests   46–7
Victoria   8, 173
Vital Signs program (US)   107
volunteer networks   35, 39, 52, 53, 94, 130, 141–7

waddy wood (*Acacia peuce*)   160
water accounting   81–4
waterbird breeding   186
waterbird surveys   26, 50, 56, 171–7
Waterwatch   144
weather events   113, 152, 160
web-based decision-making tools   202
Wentworth Group of Concerned Scientists   25, 122
Western Australia   44, 122, 123, 135, 160, 161
western Queensland   112–13, 157, 159
wet forests   198
wetlands   173, 176
Wet Tropics Management Authority   37
Wimmera   92
woodland fauna   56
workshop (ANU)   202–3
World Commission on Protected Areas (WCPA)   102, 107
World Wildlife Fund (WWF)   145

Yukon   151–5

Zerger, Andre   123

www.ingramcontent.com/pod-product-compliance
Lightning Source LLC
Chambersburg PA
CBHW041129280526
45792CB00013B/2365